T0328587

The Cost of Electricity

The Cost of Electricity

Paul Breeze

ELSEVIER

Elsevier
Radarweg 29, PO Box 211, 1000 AE Amsterdam, Netherlands
The Boulevard, Langford Lane, Kidlington, Oxford OX5 1GB, United Kingdom
50 Hampshire Street, 5th Floor, Cambridge, MA 02139, United States

Notices
Knowledge and best practice in this field are constantly changing. As new research and experience broaden our understanding, changes in research methods, professional practices, or medical treatment may become necessary.

Practitioners and researchers must always rely on their own experience and knowledge in evaluating and using any information, methods, compounds, or experiments described herein. In using such information or methods they should be mindful of their own safety and the safety of others, including parties for whom they have a professional responsibility.

To the fullest extent of the law, neither the Publisher nor the authors, contributors, or editors, assume any liability for any injury and/or damage to persons or property as a matter of products liability, negligence or otherwise, or from any use or operation of any methods, products, instructions, or ideas contained in the material herein.

Library of Congress Cataloging-in-Publication Data
A catalog record for this book is available from the Library of Congress

British Library Cataloguing-in-Publication Data
A catalogue record for this book is available from the British Library

ISBN: 978-0-12-823855-4

For information on all Elsevier publications visit our website at
https://www.elsevier.com/books-and-journals

Publisher: Joe Hayton
Acquisitions Editor: Graham Nisbet
Editorial Project Manager: Aleksandra Packowska
Production Project Manager: Prasanna Kalyanaraman
Cover Designer: Matthew Limbert

Typeset by TNQ Technologies

Contents

Chapter 1

The cost of electricity, an overview

Electricity is arguably the most important source of energy in the world. Without it, the features that make our societies modern cannot operate. From electric light bulbs to the most advanced mobile phone, modern technologies cease to function without electricity to power them.

Electricity is environmentally important too. The electricity industry is the single largest global producer of greenhouse gases through the extensive use of fossil fuel power stations. In 2018, 42% of all energy-related carbon dioxide emissions were produced by the power sector.[1] Reducing the emissions from these plants is a key to managing global warming. At the same time, electricity is a major part of the solution to this problem. Electricity generated without the production of greenhouse gases, from renewable or nuclear power plants,[2] can be used to supplant the use of fossil fuels not only for powering advanced technological equipment but also in helping to meet vital human needs for heating (or cooling) and cooking. If we want to eliminate combustion technologies that burn coal, oil, gas or wood but still maintain or improve living standards across the globe, electricity is the only viable substitute capable of meeting all these needs.

Table 1.1 shows figures for the total production of electricity across the world each year between 1999 and 2019. Over this 20-year period, total global production increased from 14,918 TWh to 27,005 TWh, an increase of 81%. Much of this increase has been fuelled by growth in Asia. The figures also show that electricity production has increased year upon year except between 2008 and 2009 when production decreased slightly. The fall coincided with the global financial crisis which affected the economic performance of most nations.

Table 1.2 shows how electricity production was broken down regionally in 2019. The largest regional production was in the Asia Pacific region with 47%

---

1. *Tracking Power 2019*, International Energy Agency, https://www.iea.org/reports/tracking-power-2019.
2. Some will argue that nuclear power plants do not belong here but that is a debate for another place. This volume is concerned solely with costs, both economic and environmental, that pertain to the production of electricity.

The Cost of Electricity. https://doi.org/10.1016/B978-0-12-823855-4.00001-2

1

TABLE 1.1 World electricity production 1999–2019.

Year	Annual electricity production (TWh)
1999	14,918
2000	15,555
2001	16,789
2002	16,345
2003	16,924
2004	17,726
2005	18,454
2006	19,155
2007	20,046
2008	20,421
2009	20,264
2010	21,570
2011	22,257
2012	22,805
2013	23,434
2014	24,030
2015	24,266
2016	24,923
2017	25,643
2018	26,653
2019	27,005

Source: BP Statistical Review of World Energy 2020.

of the aggregate total. North America, with 20% of annual production, had the second largest output followed by Europe with 15%. At the bottom of the table was Africa, which accounted for only 3% of the total. This figure suggests that compared with other regions, in Africa access to electricity is limited.

Total energy consumption from all sources across the globe was 583.9 EJ in 2019, or 162,194 TWh. Total electricity production of 27,005 TWh

TABLE 1.2 Regional electricity production in 2019.

Region	Annual production (TWh)
Asia Pacific	12,691
North America	5,426
Europe	3,993
CIS (Commonwealth of Independent States)	1,431
Central and South America	1,329
Middle East	1,265
Africa	870
World total	27,005

Source: BP Statistical Review of World Energy 2020.

represents around 17% of this, not allowing for losses between production and consumption.[3]

The pivotal role of electric power in the modern world makes the cost of electricity an important indicator and determinant for societies across the globe, both economically and environmentally. The cost of electricity, where it is available, determines who has access to the energy source. The higher the cost of a unit of electricity, the more difficult it becomes for those on lower incomes to use it freely. And, from an environmental perspective, electricity will only be able to replace other, more polluting energy sources if it is cheaper than those other sources. Otherwise, combustion fuels will continue to be employed. Our future environmental safety requires electricity to be affordable for all.

At an industry level, cost is equally important. As economies expand and require more electricity, cost will determine either wholly or in large part the type of new power plant that is built even if this overrides environmental considerations. This effect can be clearly seen in action today. Fossil fuels, especially coal, are cheap and so in spite of the environmental cost of building new coal-fired power stations, new coal-fired power stations continue to be built.

The lifetime of a power plant

It is not only expanding economies that require new power stations but also established economies. Like all modern industrial products, power stations

---

3. BP figures for global electricity consumption are not available.

have a finite lifetime. For some, hydropower stations, for example, the lifetime may be as much as 100 years. However most of the technologies that are used to generate electric power have much shorter lives than this, typically 20–30 years. This means that every 30 years or so, old power stations must be decommissioned and new ones should be built to replace them. Thirty years is a long time compared with the lifespan of many modern products such as motor vehicles or electronic devices, but even so, like those other products, the stock of power stations must be renewed regularly. When this happens, cost comes into play.

Each time a new power station is needed, a study will be carried out to determine what type of power plant will offer the best return. Many factors may be taken into account including the environmental impact of different technologies, but in most cases either the economic return that the project will offer, or its cost, will take precedent. That means that clean technologies must be able to produce electricity more cheaply than dirty technologies if they are to be preferred.

The lifetime of a power plant is also a valuable metric by which to examine aspects of its performance other than its economic ability. We can, for example, compare the environmental cost of different power plants over the lifetime of each, enabling us to make a comparison of the relative environmental impact of each type. Or we can examine the lifetime efficiency of a plant in converting the energy it exploits into electricity. Lifetime analyses of this type provide a useful means of gaining insight into other sides of performance and they can reveal unexpected benefits or deficiencies.

The main factors contributing to the cost of electricity

On purely economic terms, there are two key factors which determine the cost of a power station. The first of these is the capital cost of a power plant, which is the cost of manufacturing or constructing the components of the station and the cost of erecting them. Depending on the type of power plant, this will be a sum of material costs, the manufacturing costs and the labour costs. The capital cost of most power stations is significant and in many cases this can only be met through some form of debt financing. The cost of this financing will also feed into the final cost and so can be considered as a part of the capital cost.

The other major contributing factor is the cost of the fuel used by the power station. For combustion power plants, this will be the cost of the coal, oil, gas, wood or waste that is used to fire the plant. For a nuclear power plant, it is the cost of the nuclear fuel upon which this technology depends. In the case of generating technologies that depend on renewable energy sources, the fuel — be it water flowing along a river, the wind or the energy from the sun — is usually free.

The fact that most conventional power stations rely on a fuel that must be paid for while renewable energy is generally available without cost means that the cost effectiveness of the two types of technology depends critically on the balance between fuel cost and capital cost. For example, gas-fired power stations can be extremely cheap to build, but the fuel, while is often cheap, can sometimes become very expensive. This can make them periodically uneconomical to run. A hydropower station, on the other hand, will probably cost a lot to build but will provide power at little cost once it has been paid for.

There is a third factor that also contributes to the cost of electricity, the cost of operating and maintaining a power station. However, because all stations require this, the relative effect on the overall cost is small.

Network factors

Electricity has some of the qualities of a commodity. It can be bought and sold on national and international markets and the price is dependent in part at least on demand. However, unlike a commodity, electricity has no physical presence. It is not possible to buy or sell a barrel of electricity. Electricity is ephemeral. As soon as it is generated, it must be used. This means that the production of electricity must be carefully balanced against demand. If these come out of balance, problems can result.

For the majority of electricity users, power is delivered across an electricity network. This power is generated by an array of different power stations that are connected to the grid. At the centre of this web is a system control centre where the amount of electricity fed into the network is balanced against the amount being taken out. For this balancing act to be possible, the output of the power stations on the grid must be capable of being modulated at will.

This was traditionally achieved by using different types of power station with different characteristics, some providing constant output, the base load for the grid, others providing a variable output to meet varying demand as conditions changed. This traditional system has been upset in recent years by the introduction of large quantities of renewable energy from wind and solar power stations. The output of these plants will vary with weather conditions, so the amount available to the grid varies, adding another source of volatility to the grid balance equation. In addition to demand on a grid changing with time, supply now varies too in a way that did not happen previously. Using variable renewable energy effectively requires additional technologies to keep the grid in balance and this can add to their cost relative to the more conventional power sources.

When electricity is delivered across a network, there are additional costs added to the price of the product. The transmission and distribution of electricity across the wires of the network is not 100% efficient; it involves losses. A part of these are the electrical losses inherent in any system in which electrical power flows along wires. Other losses are due to poor maintenance or to theft. In addition, the cost of managing the network has to be taken into account. The net result is that the price of the electricity purchased by a consumer at the end of the network will be significantly higher than the cost of the electricity when it entered the network. In this book we will be dealing mostly — but not entirely — with generating costs, the cost of electricity at the point it enters the network.

Environmental and structural factors

Although there is growing awareness of the environmental impact of human activity such as power generation, the cost associated with this is still not widely taken into account when determining the actual cost of a kilowatt of electrical power. It is true that there are carbon taxes or levies of various sorts that have been introduced by different nations or regions, but these have been piecemeal and they do not yet force power generators to pay the full cost of the environmental damage they cause.

This imbalance normally favours combustion technologies at the expense of renewable technologies. Righting the imbalance requires action at a government level, but in most cases this means damaging the economic activity of the nation in question because paying the full cost of the environmental impact of emissions such as carbon dioxide will significantly increase the cost of electricity in many places. Thus there is a political dimension to electricity costs and also a political risk factor for power generators. If governments should decide to act to counter environmental damage, this might suddenly affect the economic viability of some types of power plant. This could potentially leave coal-fired power stations and others as 'stranded assets' with outstanding debts to pay but no income from which to pay them.

For the moment, that has not happened to any significant extent. But ignoring the environmental cost of a particular generating technology is not the only way that the true cost of generating electricity can be distorted. Another is through selective subsidies. Again this operates at a national level, when a government seeks to protect or encourage a particular industry by artificially lowering its costs. There has been widespread use of subsidies to encourage the construction of renewable energy power plants. This helps to make them more competitive relative to other technologies. However, most analyses show that the largest subsidies around the world are aimed at fossil fuels.

Another type of subsidy that is particularly common in oil and gas producing countries is a tariff subsidy that makes the cost of electricity artificially low. The inevitable consequence of this is that consumers use more electricity, and where this is generated by the local oil or gas, environmental emissions become elevated. Tariff subsidies are used across the world, often to alleviate poverty. However, such subsidies are often poorly targeted and, again, can lead to distortion in consumption patterns. In many cases, this type of subsidy is a political tool too.

The cost of electricity

The historical cost of electricity is known. It can be extracted from various data stores and laid out in tables such as those found in later chapters of this book. These data can often be broken down to show how different factors such as capital cost and fuel cost contribute to the final cost of energy. Costs to different types of consumer can be shown too, as well as the distribution of consumption between different sectors. These types of data will show us how prices and consumption have varied over time and it can indicate trends.

The most useful cost of electricity to know, however, is the cost of a unit of electricity at some point in the future. That is the figure that is needed when trying to determine which type of new power station is going to provide the most economical source once it has been built. That is the figure that planners and proposers of different sorts want to know. Or, it is the figure that shows how much more (or less) future electricity might cost if government policy favours a particular technology, such as wind power or nuclear power.

Some future costs can be extrapolated from historical costs. An operating power plant that provided electricity for a certain price yesterday can be predicted with some certainty to provide electricity for a closely related price tomorrow. But future planning, say when deciding what type of new power plant to build, will likely depend on knowing the cost of the electricity from different types of power plant up to 30 years hence. Such figures can only be determined with limited certainty. Estimating them depends on making guesses about a number of unknowable factors and then using these guesses to work out what the end result might be. In other words, it depends on modelling.

Modelling for the future cost of electricity is well established and widely used, and this book uses figures from such economic models. But there is a high degree of uncertainty associated with the results of such modelling and this must always be taken into account when using the figures these models produce.

This book will be mainly concerned with numbers rather than models. Its purpose is to provide as much guidance as is available from historical costs and from the historical output of economic modelling of the future cost of electricity. It will not be laying out those models or discussing them in any detail, or at least only in as much detail as is necessary to understand where the numbers come from. The data that form the heart of the book are taken from internationally credited sources and will be the most up to date at the time of publication. The aim is not to provide an economic analysis of the electricity industry. That can be found elsewhere. The aim is, rather, to assemble as much data as are available today to delineate the important cost trends that can be established. The philosophical question of whether the future can be predicted from past behaviour aside, these data can, I would suggest, provides valuable insights.

Chapter 2

The power generating technologies

Modern electricity production depends on a range of technologies that convert one form of energy into another. Before the widespread use of electrical power, several of these technologies were used to provide mechanical power.

Historically, hydropower was the first type of generating technology to emerge from these earlier devices. This technology uses a turbine to convert the energy in flowing water into rotary motion. By the 19th century, water wheels — simple turbines — had been used for millennia as sources of mechanical power, so exploiting the technology to drive a rotating dynamo and produce electricity was an obvious step. This was swiftly followed by the adaptation of combustion technology, already well known during the nineteenth century too from steam engines, to produce electrical power from oil or coal. The rapid growth of electricity systems during the twentieth century then led to the development of a diverse range of new energy conversion technologies such as solar power that had no direct precursors. Electricity production from the most important sources for the year 2017, the latest year for which complete figures are available from the International Energy Agency (IEA), is shown in Table 2.1.

Neither the historical development of electric power nor the technological niceties of power generation technologies are of importance to the understanding of the cost of electricity. What is important is to understand the different characteristics of the technologies and how these contribute to the way they are valued as sources of electrical power. For example, combustion power plants can usually be turned on or off at will and so they are considered as reliable sources of electricity. However, many renewable generating technologies depend on a variable and often unpredictable energy source. This renders them inherently less reliable. Or, some technologies are fast acting, so they can be brought into service quickly. Others have an inherent inertia that makes them slower to bring on line. Again, this will influence their perceived usefulness.

This chapter will give a brief overview of the important electricity generating (and storage) technologies highlighting the characteristics that have a bearing on energy costs.

The Cost of Electricity. https://doi.org/10.1016/B978-0-12-823855-4.00002-4

TABLE 2.1 Global electricity generation by source, 2017.

Source	Electricity generation (GWh)
Coal	9,863,339
Natural gas	5,882,825
Hydropower	4,197,299
Nuclear power	2,636,030
Wind power	1,127,319
Oil	841,878
Biofuels	481,529
Solar PV	443,554
Power from waste	114,043
Geothermal power	85,348
Other sources	36,022

Source: Electricity Information 2019, International Energy Agency.

Coal-fired power generation

Coal-fired power generation is both the most important and the most polluting type of electricity generation in use today. According to the IEA, the production of electricity from coal-fired power plants exceeded 10,000 TWh for the first time in 2018,[1,2] a significant milestone. Overall output from coal-fired plants increased by 2.6% between 2017 and 2018, and total production was equivalent to 38% of global electricity generation of 26,300 TWh.

The use of coal for electricity production is not geographically uniform. The continued growth in coal-fired power generation is concentrated in China, India and Southeast Asia. The use of coal for electricity generation, meanwhile, dropped in the United States and Europe between 2017 and 2018, a trend that is expected to continue as renewable sources become increasingly important to these regions.

Coal-fired power stations are relatively expensive to build, but the fuel they burn, coal, is cheap. This makes coal the fuel of choice in many countries that have coal deposits. However, the fuel is costly to transport over great distances, so while there is an international market for high-quality coal, this is

---

1. *Tracking Power 2019*, International Energy Agency, https://www.iea.org/reports/tracking-power-2019.
2. *Tracking Power 2019*, International Energy Agency, https://www.iea.org/reports/tracking-power-2019.

relatively limited with only around 21% being traded internationally in 2018 according to the World Coal Association.[3]

The most common technology for coal-fired power generation involves burning pulverised coal in air in a specially designed boiler where the heat generated is used to raise steam, and the steam is used to drive a steam turbine. The best steam turbine–based coal plants can achieve an energy conversion efficiency of around 45%, but many older plants are much less efficient. The efficiency of the Japanese fleet of coal-fired plants in 2016 was around 42% and that of the US fleet was 34% according to the IEA. Meanwhile, the IEA has put global coal plant efficiency at the end of the second decade of the 21st century to be around 37.5%.[4]

When coal is burnt in air, the major product of the combustion process is carbon dioxide. The more efficient a plant is, the less carbon dioxide it produces for each unit of electricity it generates, so high-efficiency plants are considered cleaner. However, the technology for coal-fired power plants is already highly optimised and the only way that efficiency can be increased is by increasing the operating temperatures and pressures of the steam in the plant. This puts a heavy load on the steam plant components, many of which need to be made from specialised materials to be able to withstand the conditions. Advances in material development are slow and new materials are often costly.

Coal combustion is the most significant source of electricity sector carbon dioxide, but it is possible to capture the gas that is produced in a coal-fired power station before it enters the atmosphere. Various technologies capable of achieving this have been developed. The use of these technologies reduces the efficiency of the power plant, thereby increasing the overall cost of the electricity it produces. The carbon capture technology also increases the overall capital cost of a coal plant.

In addition to carbon dioxide, coal combustion produces a range of other harmful emissions including sulphur dioxide, nitrogen oxides, heavy metals and small dust particles. All these harmful emissions must be captured before the flue gases can be released into the atmosphere. Again, these emission control features increase the capital cost of the power generating facility.

Modern coal-fired power stations are generally large with single units up to 1000 MW and power plants of multiple units with capacities of 5000 MW, sometimes more. The largest plants are often built close to the mines that supply them with fuel.

A coal-fired power station has a relatively high mechanical inertia making it slow to start and these stations have traditionally been used to provide base load power. More recently, coal plants in some regions have had to

3. World Coal Association, https://www.worldcoal.org/coal/coal-market-pricing.
4. *Historic Efficiency Improvement of the Coal Power Fleet*, Qian Zhu, International Energy Agency, 2020.

provide a much more variable output, to the extent that in the United Kingdom in 2020, coal plants have been used as peak load plants. However, the plants are not easily or economically adapted for this type of service.

While the high inertia of a coal plant makes it less flexible than some technologies when it comes to variable output, large inertia can be advantageous in other situations. The massive steam turbines in a coal plant have a high rotational momentum when spinning and this enables them to be used to help stabilise fluctuations in the grid caused by variations in supply and demand elsewhere. This 'spinning reserve' is a valuable resource for grid stability, particularly as more renewable plants with variable output are connected to the grid. However, the polluting nature of coal makes it unlikely that coal stations will be maintained simply to provide this spinning reserve.

Coal-fired power generation without carbon capture remains one of the cheapest forms of power generation and this has so far enabled coal-based generation to continue to flourish in spite of its high environmental cost. However, the most modern renewable sources are now challenging it on cost, if not yet on capacity. Unabated[5] coal-fired generation must decline in the coming decades if global warming is to be controlled. The additional cost of carbon capture makes plants with this technology much less attractive economically and it is likely only to be used as a transitional technology while alternative technologies such as wind power and solar power are built up and integrated into grids.

In spite of this, the economics of coal has made many nations reluctant to abandon coal-fired generation. This remains one of the most contentious areas of the global warming debate.

Gas-fired power generation

Natural gas—fired power plants are the second most significant category of fossil fuel power plants in operation today. During 2018, power plants that exploited this fuel generated 6100 TWh of power, an increase of 4% over the figure for 2017 according to the IEA.[6] This was pushed by strong growth in the United States where production rose by 17% and China where overall production rose by 30% but from a relative low base. In contrast, production of power from natural gas fell by 7% in Europe where renewable generation is advancing strongly at the expense of conventional sources. The fuel accounted for 23% of total global power generation in 2018, well below that from coal-fired plants.

---

5. Unabated is the term commonly used to describe a coal plant without any carbon dioxide capture.
6. *Tracking Power 2019*, International Energy Agency, https://www.iea.org/reports/tracking-power-2019.

Unlike coal, natural gas is traded both nationally and internationally and this has led to widespread use. While reserves of the fuel can be found in many regions, a small number of countries hold the majority of known reserves. The Russian Federation, Turkmenistan, Iran and Qatar held 58% of total global reserves at the end of 2017 according to the BP Statistical Review of World Energy.[7] Other major producers include Saudi Arabia and the United States.

Natural gas is usually distributed over land using natural gas pipelines. These can be national systems as in the United States and in many countries in Europe, or they can be transnational. Much of Europe's natural gas comes through pipelines from the Russian Federation. Further international trade in natural gas is carried out using liquefied natural gas which can be stored in container vessels and shipped around the world. This is then off-loaded at natural gas terminals before being fed into local pipeline systems.

Natural gas can be burnt in a boiler in the same way as pulverised coal to produce steam that drives a steam turbine to produced electricity. However, the most important means of exploiting natural gas for power generation is with a gas turbine. These devices use natural gas combustion to generate a stream of hot, high-pressure air that is then used to drive an air turbine to generate rotary motion which powers a generator.

Simple gas turbine power plants can have efficiencies similar to that of the best coal-fired power plants at around 46% energy conversion efficiency. Higher efficiency can be achieved by using a more complex configuration called a combined cycle plant. This uses a gas turbine as the primary turbine generator. The exhaust gases from the gas turbine, still at high temperature, are then used to raise steam in a waste heat boiler and this steam drives a steam turbine. This configuration, using large gas and steam turbines, can reach an energy conversion efficiency of perhaps 61% in the best plants today.

The combustion of natural gas produces a combination of water (as water vapour) and carbon dioxide. There may be a small proportion of other impurities such as unburnt hydrocarbons and carbon monoxide. The combustion process can also produce significant quantities of nitrogen oxides from air as a result of the high temperatures reached. The nitrogen oxides and other minor impurities are removed in gas turbine power plants, but there remains a significant amount of carbon dioxide. However, the quantity produced during combustion is much lower than for the combustion of a similar quantity (in energy terms) of coal. This, together with the higher efficiency, means that the carbon intensity[8] of a natural gas−fired power plant is much lower than that of a coal-fired power plant.

Natural gas turbines come in many sizes, from small units of a few megawatts to massive turbines with generating capacities of 600 MW. The

7. *BP Statistical Review of World Energy 2019.*
8. The carbon intensity is the amount of carbon dioxide released into the atmosphere for each unit of power delivered to the grid.

very large gas turbines are generally designed exclusively for combined cycle plants; a single combined cycle unit may be capable of generating up to 800 MW of power, comparable to the largest steam turbines in coal plants. However, unlike coal plants, gas turbine combined cycle plants are relatively flexible and can be adapted easily to grid support and peak power production.

Traditionally, gas turbines have been used for a range of grid services. The largest combined cycle plants were intended for base load operation while small, simple cycle gas turbines (those without attached steam turbines) were used for producing power at times of peak demand. These small units, based on aero engines, can be stopped and started very quickly. However, the electricity from a small, peaking gas turbine is likely to cost much more than from a large combined cycle gas turbine plant.

Gas turbine power plants are among the cheapest power plants to erect. In consequence, the electricity from these large plants can be extremely cheap. However, the cost depends critically on the natural gas price which can be highly volatile. In times of high natural gas prices, many combined cycle power plants stand idle because they are not economical to operate.

The lower atmospheric carbon dioxide emissions from gas turbine power plants have led to these being used in many countries and regions to reduce emissions in order to meet global warming targets. Where these plants are displacing coal-fired generation, this will lead to a reduction in national emissions. However, substitution of natural gas for coal can only serve as a temporary measure as they still produce significant quantities of carbon dioxide. As with coal plants, it would be possible to capture the carbon dioxide from natural gas production, but this makes the plants less economical to operate. Over the longer term, therefore, it is likely that many natural gas-fired plants will be replaced by renewable energy sources. However, natural gas plants may have a longer role to play in grid support where their ability to respond rapidly to demand changes makes them a good match for renewable energy.

Piston engine−based power generation

Piston engine power plants encompass a diverse range of energy conversion devices that all exploit the movement of a piston in a barrel as the primary means of converting heat energy in kinetic energy. They include diesel engines and spark ignition engines − both types of internal combustion engine − as well as external combustion engines such as the Stirling engine. The size of units employed for electricity generation can vary from 1 kW to 65 MW. Efficiencies vary widely too. Virtually all internal combustion engines used for power production burn fossil fuels and so they produce carbon dioxide as well as a range of other pollutants. External combustion engines can exploit energy from other sources, too, such as the sun.

With such a disparate array of devices falling into this category, it is difficult to assess their global contribution to electricity production. However, estimates suggest that perhaps 50–60 GW of generating capacity based on these engines are installed every year.

The majority of the piston engines that are used for power generation are derived from engines developed initially for transportation applications. The smallest engines in use are often based on automobile engines. These engines are cheap, inefficient and do not last for very long (in power plant terms). This is a disadvantage where an engine is required for continuous duty but many of these small engines are used in power backup systems where they are only required to run under emergency conditions so that short operating life is unimportant. These engines usually operate on petrol or diesel.

Larger engines, in 10 kW–5 MW range, are variously derived from truck or railroad engines. These engines have much longer operating lives and when adapted for power generation service can last as long as other fossil fuel generating technologies. They are also much more efficient than the small engines. They comprise both spark ignition and compression ignition (diesel) engines. The former have found a particular use in burning natural gas. Gas engines, as these are often known, are relatively clean compared to large diesel engines. However, the latter can be much more efficient; a large diesel engine may be capable of an efficiency of 48%, whereas the efficiency of a gas typical spark ignition engine is around 40%.

The largest engines of all are based on marine engines. These can be as large as 65 MW. They operate at very slow speeds compared to smaller piston engines and they can burn very poor fuels. In addition, it is possible to configure these large engines in a combined cycle mode by adding a small steam turbine. With this, energy conversion efficiency can be in excess of 50%.

Applications of these engines are as various as their sizes. Many are used in emergency backup systems and medium-sized gas engines are popular providing power to municipal facilities such as in hospitals as they are relatively clean to operate. Most piston engines are capable of load following and their efficiency will often barely fall when output falls from 100% to 50%. This can make them attractive for providing peak power on grid systems. The largest engines are normally used for base or intermediate load on a grid. Piston engine power plants have also been widely used to provide power to remote communities that are unable to be connected to a grid.

Another important use for small- and medium-sized piston engines is in cogeneration systems. In these systems, waste heat from the energy conversion process is captured and used to provide hot water, or in some cases to provide heat for an industrial process. Since most engines have efficiencies of well below 50%, more than half of the input energy is wasted. Capturing are using the waste heat increases overall efficiency significantly. However, the application requires a local heat demand such as a hospital to make cogeneration practical. Cogeneration can also be used domestically.

Stirling engines are a novel form of piston engine in which the heat energy to drive the cycle is applied externally. This allows them to be used in a range of applications, but the most significant is for solar energy conversion when the sun is used as a heat source.

Emissions from piston engine power plants depend upon the precise fuel. Technologies for removing nitrogen oxides and for reducing the emissions of particles from diesel engine exhaust gases are widely deployed. However, all these engines, when they burn a fossil fuel, will generate significant quantities of carbon dioxide. It is unlikely ever to be economical to remove carbon dioxide from the exhaust of this type of power plant.

Hydropower

Hydropower is the most important renewable electricity generating technology and the earliest renewable technology to provide a significant part of global power generation. In 2017, hydropower plants generated 15.9% of total global electricity production according to the IEA.[9] The aggregate global installed capacity was 1270 GW in 2016 and the figure increases, year upon year, but the rate of increase is lower than the rate of increase in total generating capacity so that the proportion of power generated from this source is declining.

Hydropower potential is found in most parts of the world, excepting the most arid, where water flows in streams and rivers to the world's oceans. The most developed regions and nations, such as the United States and Europe, have exploited the best of the potential available, but elsewhere there is still significant hydropower potential that could be exploited. Africa, in particular, could generate significant volumes of electrical power from its resources. Against this, the development of large hydropower schemes can be extremely disruptive environmentally, so great care is needed when new projects are developed.

Hydropower is a variable energy resource. The amount of energy available changes by the season and depends on annual rainfall. Thus, the annual availability can vary significantly. Global warming, which is changing weather and rainfall patterns, can also have a major effect on availability, and over this century it is likely that availability in some regions will fall while in others it will rise.

Hydropower plants can vary in size from a few kilowatts to tens of thousands of megawatts. Sites capable of providing sufficient flowing water for the largest projects are rare but plants that range in size from tens of megawatts up to a thousand megawatts are relatively common. There are two primary types of hydropower development, projects that involve building a dam and reservoir and projects that do not have a reservoir; this latter is usually called a

---

9. *Key World Energy Statistics*, International Energy Agency, 2019.

run-of-river project. Hydro project reservoirs can often cover large areas of land, displacing wildlife and people. Against this, they provide a means of energy storage because they store water during the wettest seasons and can make that water available for electricity generation all year around. The largest projects often provide water for irrigation as well as other amenities, helping to improve living standards locally. In consequence, the largest of projects are often funded nationally or through international lending agencies.

Smaller projects are more often commercial although they may still be massive in scale. Most projects of 1 MW or more in generating capacity will be connected to the regional grid to provide electrical power to a region or nation. Smaller projects in the 10−1000 kW range might provide their power locally to a small community or to an industrial facility. Smaller projects still are built to provide power to a single dwelling, often in a remote region where there is no grid power available.

The nature of a hydropower project, particularly one that involves the construction of a dam, is such that the capital cost of its construction will be high. However, the major parts of such a scheme, the dam, waterways and the powerhouse will have a long lifetime. Provided the turbines are maintained and repaired regularly, a large hydropower scheme can operate for a century or more. This means that once the project has been constructed and the cost has been met, the plant will produce extremely cheap power. Smaller plants − and schemes without dams and reservoirs − are cheaper to build. The unit cost of a small hydropower scheme is usually higher than that of a large scheme.

Hydropower is a renewable resource and this has made it important as global warming had threatened the world. The output of a hydropower plant is usually predictable, but it can vary significantly seasonally. In consequence, hydropower alone cannot be relied upon as a secure source. From a grid perspective, hydropower plants can be brought online and taken offline rapidly and this can make them valuable for grid management. In addition, the large turbines in some of the biggest hydropower schemes can provide spinning reserve to aid grid stability.

One of the most important roles of hydropower for modern grid management is its ability to store energy. A hydropower scheme that includes a reservoir has a natural store of energy in the form of the water behind the dam. This type of plant can be used to help balance a grid, providing power when other renewable sources such as wind and solar cannot and backing off when the output from these plants is copious. There is a specific type of energy storage plant that exploits this concept called a pumped-storage hydropower plant, but any plant with reservoir storage can provide this service so long as water remains in its reservoir. Grids with large amounts of hydropower are therefore easier to manage than those without and the electricity they provide is often cheaper.

While hydropower plants have many benefits, they can also lead to environmental problems. As already noted, the displacement resulting from a large hydropower scheme can be extremely damaging if not managed well. Furthermore, hydropower reservoirs are capable of producing large quantities of methane, a potent greenhouse gas, during the early years of their establishment. A dam can disrupt migratory fish movements and the movement of silt downstream, this latter often important for downstream land fertility.

There is one other type of hydropower plant that can potentially provide significant quantities of clean energy and that is the tidal power plant. Tidal power plants use the same types of technology as conventional hydropower schemes, but instead of taking energy from flowing rivers, they rely on the ebb and flow of tidal waters. Tidal power is expensive to develop but could potentially offer a massive amount of energy. In addition, tidal power is entirely predictable making grid management of the output much simpler than for other hydropower schemes.

Nuclear power

Nuclear power is perhaps the most contentious of all the power generation technologies in widespread use today. The technology has obvious attractions because a nuclear power plant does not generate carbon dioxide during the energy conversion process it exploits. Against this the potential dangers of nuclear power, both because of its link to nuclear weapons technology and because of the environmental damage that can be caused if a nuclear power plant fails, mean that while some nations continue to embrace nuclear technology, others are proposing to reduce its presence or abandon it completely.

Nuclear power plants accounted for 10.3% of global electricity production in 2017 according to the IEA[10] ranking the technology as the fourth most important in terms of output after coal, natural gas and hydropower. Total global capacity that year was 392 GW, the largest part of which, 100 GW, was located in the United States. Other major nuclear nations include France, Japan, China, the Russian Federation and South Korea. However, global capacity has been relatively flat since 2000 with only small capacity additions. Furthermore, many of the power plants in operation are ageing and without replacement or remedial action will have to be retired from service over the next decade or two.

Nuclear power exploits the ability of the atoms of certain large natural elements to split into smaller atoms with the release of large amounts of energy. These reactions are at the heart of nuclear weapons, but the same process, if allowed to take place under controlled conditions, can be used to make a power plant. In this case the nuclear reactor, in which the nuclear process takes place, acts essentially like the combustion boiler in a coal-fired power station, producing heat that can be used to raise steam and drive a steam turbine.

---

10. *Key World Energy Statistics*, International Energy Agency, 2019.

Nuclear steam cycles are relatively inefficient compared to those in coal-fired power plants. An overall heat energy to electricity conversion efficiency of 33% is typical.

The nature of the nuclear reaction makes the safety of nuclear reactors of paramount concern. A reactor that became out of control would in essence be a nuclear bomb. Reactors are therefore complex, high technology installations that include multiple safety features to try to ensure that the facility can never fail. Nuclear reactors are therefore expensive to build, probably the most expensive type of large-scale power plant in common use. In contrast to the high capital cost, the cost of the fuel for a nuclear power plant is relatively low and this enables them to compete with fossil fuel power plants. However, there are a large number of additional costs such as spent fuel reprocessing and power plant decommissioning that can elevate the overall cost of power from these power stations.

Most of the nuclear power stations in operation today are relatively large with single reactor sizes often in excess of 1000 MW; there can be several of these on one site. The capital outlay for such a project can be intimidating. There is a great deal of interest today in small nuclear reactors, units of much more modest generating capacity which can be built in factories and then shipped to a site. This might reduce the cost of nuclear power significantly. In addition some of the designs being developed operate with high steam cycle temperatures, leading to higher energy conversion efficiency.

There is another type of nuclear reaction called nuclear fission that involves the reaction between the atoms of small elements to create atoms of a larger element, again with release of large amounts of energy. This nuclear fission reaction has the potential to provide large amounts of electricity relatively cheaply if it can be developed to a level to make it available commercially. However, in spite of decades of work, a commercial fusion reactor still remains a long way off.

Conventional nuclear reactors of the type most common today were originally designed to be operated as base load power plants, operating at full power for very long periods without interruption. This remains the duty cycle of many plants but as with other large power plants, nuclear plants today are also being expected to modulate their output in order to cater for variable amounts of renewable energy on a grid. One strategy used in the past to manage nuclear plant output has involved building very large energy storage plants based on hydropower to store surplus energy from the nuclear generating facility when it is not needed, then making this energy available as demand peaks. This is a capital-intensive strategy.

Nuclear power plants are attractive to many nations as a means of ensuring a secure electric power supply that is independent of the vagaries of the fossil fuel market. Countries such as the United States, France and Japan have invested heavily in nuclear technology for this reason. However, there are questions in all these countries, and in others, about the environmental safety of their nuclear facilities. Some nations categorise nuclear power as renewable energy source but most environmentalists would challenge this. There are also major differences of opinion about the economic effectiveness of nuclear power.

Meanwhile the development of nuclear nuclear facilities has begun to accelerate, slowly. According to the IEA, 11.2 GW of nuclear nuclear capacity was brought online in 2018, the largest amount since 1989. Only a very small number of these new plants are in developed countries that already have nuclear fleets but some of these nations are carrying out work to extend the life of their existing plants, many of which are 30 or more years old. In the United States, most of the existing nuclear reactors now have permits to operate for 60 years.

Solar power

Solar power is potentially the most important renewable source of energy available to the world and solar energy the largest natural source of energy that we have to exploit. However its development as a generating technology has only become significant during the current century and global capacity remains relatively small. In 2018 the total amount of power from solar cells, the most significant means of turning sunlight into electricity, was 585 TWh according to the IEA and accounted for 2% of global electricity generation. Meanwhile the European Photovoltaic Industry Association estimated the total global installed capacity in 2018 to be 517 GW, rising to 634 GW at the end of 2019. In addition to generation based on solar cells there is also another category of solar technology called solar thermal power generation; this provided roughly 300 GWh of power in 2018.[11]

Solar energy is available in every part of the globe, but the absolute annual amount will vary significantly from place to place. In general there is more sunlight to be found near the equator and less near the poles. The amount of energy at any point on the globe is intermittent, following a daily cycle and there is a seasonal cycle too. Changes in weather also affect the amount of sunlight available, making it in part unpredictable. Solar cells can harvest sunlight in most parts of the world provided there is sunshine available. However solar thermal power plants require a reliable source of high intensity sunlight to operate efficiently. These plants are most suited to regions of high insolation. Arid parts of the world and desert regions will often provide a good resource for this type of technology.

Solar cells are easily scalable so that it is possible to have installations with a few kW of generating capacity and others with hundreds of megawatts, all based on the same technology. The wide-spread availability of the energy

---

11. This figure is an estimate based on a graph on the IEA website and should be considered approximate.

source makes is practical to harvest solar energy using solar cell arrays on the rooftops of virtually all types of buildings, from domestic dwellings to large industrial facilities. Depending on its size and situation, this type of solar installation might provide power directly to a single household or commercial facility or be connected directly the local distribution network. Very large solar cell power stations with tens to hundreds of megawatts of installed generating capacity will normally be designed to supply power to the main grid. Solar thermal power plants are not so easily scalable and most economic power plants of this type are relatively large plants intended to supply power to the grid.

Solar cells (often called solar photovoltaic devices) are solid state devices constructed using technology similar to that employed for microchips. Most are made from silicon although some other materials are also used. The manufacture of the cells requires high technology facilities and growth in global solar cell capacity is constrained by the quantity of solar cells that the world's factories can produce. In 2018, according to the IEA, 97 GW of new capacity was installed. Solar cells can exploit both direct sunlight and diffuse sunlight so that they can still operate when conditions are cloudy. However, their output will be directly related to the amount of sunlight they receive so output will fall to zero during hours of darkness. This means that grid connected solar cells require some form of support at night. On the other hand solar cell power output usually correlates closely with daytime temperature making it a good match for air-conditioning demand in hot climates.

Solar thermal power plants use mirrors to collect direct sunlight and then focus it onto a heat collector where the concentrated heat energy is used to generate electrical power. Some solar thermal plants use a cycle similar to a coal-fired plant with heat producing steam to drive a steam turbine. Others using Stirling engines to convert the heat into electrical power directly. An important difference between solar cells and solar thermal power plants is that some of the latter can store thermal energy which can then be used to generate electricity when no sunlight is available. This can make them much more reliable sources of power than solar cells and therefore much easier for grid controllers to dispatch.

The cost of solar cells has fallen dramatically over the last 20 years and at the end of the second decade of the 21st century solar cell power facilities were capable of competing on cost with most other sources of electricity. The nature of the manufacturing process means that costs are likely to drop further as manufacturing volumes increase. Solar thermal power plants are relatively more expensive but their costs are falling too. However, they are not as competitive. Nevertheless the technology has proved attractive in some arid regions, particularly when plants provide storage capability.

Solar energy is one of the important renewable energy sources, arguably the most important. It has the potential to provide a large part of global energy demand in the future. Solar cells require a lot of energy to manufacture but the energy requirement is decreasing. The lifetime of solar cells is usually expected to be around 20 years but with regular maintenance they can last much longer without degradation. Solar thermal power plants exploit more traditional power plant technologies and their costs and lifetimes are closely related to those of fossil fuel power plants. Both types of solar generator are expected to play an important role in future energy supply but solar cells will dominate. However, neither can provide a secure supply alone and so must sit alongside other generating or storage technologies.

Wind energy

Wind energy is the second most significant renewable technology after hydropower in terms of electricity production. Global output from onshore wind turbines in 2019, according to the IEA, was 1202 TWh while offshore wind farms provided a further 66 TWh, for a total of 1268 TWh. Meanwhile figures from the Global Wind Energy Council (GWEC)[12] indicate that total installed capacity for wind energy in 2018 was 591 GW, of which 568 GW was onshore and 23 GW was offshore. The total capacity rose to 651 GW at the end of 2019.

Wind energy, the energy contained in a mass of moving air, is available in most parts of the world but the size of the resource will vary from place to place depending on the wind regime. Wind energy can be harvested on land and at sea. The offshore resource is generally the most consistent, the most reliable and able to supply the highest energy intensity. Onshore wind resources are more variable because the wind must travel over a land mass and it will be affected by the contours of the land and by the ground cover. However all wind is dependent on the prevailing weather conditions and this leads to considerable variations in availability. Sometimes the wind blows intensely and sometimes it does not blow at all. This means that wind power is probably the most variable and the most unpredictable of all the renewable energy sources. Wind output reliability can be improved by coupling wind farms that are widely spaced geographically, in effect averaging output over a large area. Even so it is still possible for a whole region to become becalmed at times. Wind energy must therefore be supported by other forms of generation or by energy storage in order for it to provide a manageable resource. Wind and solar power can be complementary since the wind blows more strongly during winter while solar power is most intense during the summer. The management of wind output is one of the most challenging aspects of grid management today.

Wind energy is captured by wind turbines. When the wind industry was young, in the 1980s and early 1990s, there were a variety of wind turbine

12. *Global Wind Report 2019*, Global Wind Energy Council, 2020.

designs in use but the range has gradually narrowed so that today the market is dominated by a single type, the three blade horizontal axis wind turbine, with the turbine and its generator sitting on top of a tall tower. Wind strength increases with height so the higher the tower, the more energy can be collected at any given site. The sophistication and reliability of wind turbines has increased enormously since the pioneer days and new wind turbines can be expected to deliver power over a lifetime similar to that of other types of power generation.

There are two branches or families of wind turbines, onshore turbines and offshore turbines. Today the differences between the machines used for each are slight. Most significant is size, with offshore turbines tending to be larger than those used onshore. This is partly a matter of practicality. Transporting and erecting a very large turbine onshore can be very challenging in many locations whereas the are no limits offshore, provided only that vessels are available that can carry and install them. However, installation of turbines offshore is much more difficult than onshore and more costly. It is therefore more cost effective to install the largest turbine possible at an offshore site. Typical onshore wind turbines have generating capacities of up to 4 MW. Offshore, 6–8 MW is more typical of the capacity range, while turbines with generating capacities of 10–12 MW are expected to enter the market for the beginning of the third decade of the century. The main market for offshore wind is in European waters but China has been expanding its offshore wind capacity in recent years too.

The cost of wind energy has fallen dramatically over the last decade and over the 5 years to the end of 2019, the cost of both onshore and offshore wind had fallen on average by more than 50%, according to the GWEC. This has made onshore wind generation easily competitive in terms of cost with fossil fuel generating technologies and offshore wind is likely to be in the same position in the near future. Unfortunately the unpredictability of wind power often still leaves it at a disadvantage. One potential means of remedying this is to combine wind power with some form of energy storage. This will increase the overall capital cost of a facility but by increasing its reliability, makes the energy it produces more valuable. Various schemes are being explored including using offshore wind power to produce hydrogen which can then be shipped ashore and used as a green energy source.

The green environmental credentials of wind power make it attractive as a means of combatting global warming and most countries are building up wind capacity, some faster than others. However, it is not entirely benign. Onshore wind turbines are large additions to any landscape and they are not always welcomed by their human neighbours. This can be problematic when obtaining permits to construct wind farms onshore. There has also been an issue in the past with the danger of wind turbines to birds. However, the slow rotational speed of large modern wind turbines makes this less of a problem today. Noise, too, can be a problem onshore so it is not usually possible to erect wind turbines close to dwellings. Onshore construction is less of a problem in countries such as the United States and China where are wide expanses of uninhabited territory that can be used for wind generation. Offshore wind experiences few problems in this respect.

Biomass power generation

Biomass makes a relatively small contribution to global electricity generation, with production of 546 TWh in 2018 according to the IEA. This figure includes production from power from waste plants, so the output from dedicated biomass power plants is likely to be under 500 TWh. However, biomass remains a very important source of energy for many communities around the world and accounts for around 10% of global energy consumption. Most of this is in the form of wood burnt for cooking and heating. Power plants that burn biomass to generate electricity are relatively uncommon. Estimates for the total global generating capacity at the end of the first decade of the 21st century vary between 80 and 120 GW. A large part of this capacity is in Europe.

Biomass fuel is combustion fuel that has been grown rather than mined from the ground. It is the product of plants and trees that convert water and carbon dioxide into organic material with the aid of sunlight. Since the latter is the driving force for the photosynthesis process in plants, biomass energy might be considered as another form of solar energy. In principle all plants and trees can be used as fuel but in practice only a limited part of global biomass is useful. The most important sources of biomass fuel are biomass wastes and energy crops. Biomass pellets, traded internationally, are derived from energy crops.

Biomass wastes cover a range of materials the most important of which are agricultural wastes from the harvesting of various crops such as cereals, rice, sugar cane and maize. Wood waste from forestry management can also be exploited but this is more labour intensive to collect and therefore tends to be more costly. One important specialist category of waste is that produced by sawmills and paper plants. This waste is often used at the site to produce heat and electricity to power the industrial installation. Another specialist agricultural waste, the slurry from animal farms, is sometimes used to feed a digester which produces methane gas.

Biomass crops are specific fast-growing species that can be harvested regularly and then converted into a form suitable for burning in a combustion plant. The most common of these are prairie grasses and tree species such as willow and hybrid poplar. Grasses can be harvested annually in the autumn when they have died back and dried. The harvested material is often turned into briquettes before being sold to power plant operators. Woods cannot be harvested so frequently but with careful managment plantations can be rotated to provide a regular supply of combustible material. Harvested wood usually needs drying before use. Some power plants can burn cut wood directly but much of it is converted into pellets that can easily be shipped to power stations. There is a small but growing international trade in wood pellets from power generation.

The technologies used to burn biomass fuels to produce electricity are essentially the same as those used to burn coal. Fuel is prepared, then fed into a boiler where combustion takes place and heat is captured and used to raise steam that drives a steam turbine. Biomass fuels have a lower energy content that coal and burn at lower temperatures, so the efficiency of most biomass combustion power plants is relatively low. This is compounded by the fact that many of these plants are small, typically no larger that 50 MW. Since these are combustion plants, emission control systems may be needed to remove pollutants from the flue gases before they are released into the atmosphere.

There is another way of burning biomass fuels called co-firing that offers a more efficient means of converting biomass into electricity. Co-firing involves adding a proportion of biomass, often in the form of pellets, to the coal that is used in a coal-fired power station. The boiler in a large coal plant is much more efficient at extracting energy from fuel that that of a traditional dedicated biomass plant, so more energy from the biomass is converted into electricity. Today some coal-fired power stations in developed countries where coal is being phased out as a combustion fuel are converting to 100% biomass combustion for power generation.

The combustion of biomass to generate electricity is considered to be renewable, but only if some specific conditions are met. The argument that biomass is a renewable energy source relies on the continuous harvesting and regrowth of biomass. When a biomass fuel is harvested and burned, it generates carbon dioxide during the combustion process in the same way as coal or natural gas. However, if the same amount of biofuel is regrown, it will absorb all this carbon dioxide from the atmosphere again so the net release will be zero. While it may appear simple to maintain this cycle when power generation and the growth of the energy crop are closely coupled, there have been questions raised about the sustainability of internationally traded biomass pellets.

There is an additional worry about fuel crops, that they might be grown in place of food crops. This can be a danger if the energy crop is more valuable than a food crop, particularly if the crop comes from an undeveloped region of the world where food is scarce. Again it is the international trade in the energy fuel that is likely to lead to abuse.

There is potentially one extremely positive adaptation of biomass combustion. Since the combustion of biomass fuel is, in theory, carbon neutral, if the carbon dioxide produced during combustion is captured from the exhaust gases of a biomass plant and sequestered so that it cannot return to the atmosphere, the process will actually remove carbon dioxide from the atmosphere. This can only be cost effective in a very large combustion power plant and since the amount of biomass that can be used globally as fuel is limited, the actual reduction in atmospheric carbon dioxide that can be achieved in this way is likely to be very small. However this has not stopped some companies pursuing this goal.

The cost of electricity from biomass is likely to be higher than from a conventional coal-fired power station. However if external costs such as those associated with carbon dioxide emissions are taken into account, and if the price for emitting carbon dioxide into the atmosphere is high enough, a biomass power plant can prove to be cost-effective in comparison with coal or natural gas.

Geothermal energy

Geothermal energy is energy extracted from the earth. The core of the earth is extremely hot and this heat slowly radiates towards the surface so substrata of our planet are warmer than the surface. The temperature gradient is relatively low but there are areas of the earth where heat from deep within the planet has warmed reservoirs of water close to the surface. If this hot brine[13] is extracted it can be used to raise steam and drive a steam turbine to generate power. There are only a limited number of places around the world where underground hot reservoirs are accessible and many of those that are within reach provide only low temperature fluid, suitable for heating but not power generation. However there are a few with higher temperature fluids.

It is also possible to create a geothermal hot water source artificially. Since deep underground strata can reach very high temperatures, it is possible to drill a deep well into hot rock and pump high pressure water into the well, then extract the heated water from a second well to provide a source of energy. This technology, similar in concept to fracking to extract oil and gas from rock, has been demonstrated but is costly and has not yet been exploited commercially.

The global production of electricity from geothermal power plants in 2018 was 90 TWh according to the IEA. Meanwhile the aggregate global geothermal generating capacity in July 2019 was 14,900 MW.[14] The IEA has estimated annual capacity additions to be around 500 MW over the past 5 years. With the limited conventional resource of underground reservoirs, geothermal power generation can only make a modest contribution to global production.

The primary attraction of geothermal power generation is cost. The resource — hot underground brine — does not cost anything, although drilling down to the underground reservoir can be expensive, particularly as the location of new reservoirs can be difficult to identify from the surface. The

---

13. The water in hot underground reservoirs contains substantial quantities of dissolved mineral salts.

14. ThinkGeoEnergy, https://www.thinkgeoenergy.com/global-geothermal-capacity-reaches-14900-mw-new-top10-ranking/.

technology used to generate electricity from a geothermal reservoir is conventional, based on the steam turbine cycle. A very small number of underground reservoirs will produce live steam when drilled and this can be used to drive a steam turbine directly. More normally a very hot brine is extracted. This hot brine can be expanded into a low-pressure chamber, producing steam to drive a steam turbine, or it can be used to heat water or another thermodynamic fluid, the vapour from this second fluid then driving the turbine. Whatever the process, the spent brine is an environmental hazard and will normally be reinjected into the reservoir. Underground reservoirs are continuously heated from within the earth but the size of each reservoir is limited. If heat is extracted more quickly than heat is added, then the temperature of the reservoir will fall and it will become depleted. The same applies if too much liquid is extracted and not returned.

Geothermal power is not exactly *renewable* because energy is being taken from the earth and not replenished, but in practice the amount of energy extracted is tiny and will have no effect on the temperature of the core. A greater problem is associated with carbon dioxide that is often found alongside hot brine in underground reservoirs and which is released into the atmosphere when the reservoir is mined for energy. While the quantity involved is relatively small, perhaps 10% of the amount produced by a coal-fired power plant, it is still significant.

Geothermal power plants are technically relatively simple, and the capital cost, while significant, is outweighed by the fact that the energy source has no cost. This makes geothermal electricity generation competitive with most alternatives.

Others types of generating technology: marine power, power from waste and fuel cells

In addition to the various technologies outlined in the preceeding sections, there are three other types that are used for electricity generation: marine power generation, power from waste and fuel cells. None makes a major contribution to electricity production today, but both marine technologies and fuel cells are seen a potentially significant technologies for the future. Power from waste, meanwhile, has a small but potentially important part to play in keeping the planet clean.

The term marine generating technology encompasses a diverse array of different methods of producing electricity that have in common the exploitation of energy contained within the world's oceans. Two of these, marine current and wave power generation, have been developed to the pre-commercial stage. Ocean thermal energy technology (OTEC) has yet to show its commercial viability.

Marine current energy production uses what is essentially an underground wind turbine to take energy from moving water. The marine current devices have rotors that turn in a flowing current, providing the motive force to drive a generator. The energy density of moving water is much higher than that of moving air and a marine current turbine will be much smaller than a wind turbine for the same power output. An underwater turbine can be mounted on the sea (or river) bed or it can be deployed from a floating platform. They are generally deployed as coastal devices where they exploit the flows of water that are generated as the tide ebbs and flows. However, floating generators might also be used in the future to tap the great ocean currents such as the Gulf Stream.

Wave power generators try to capture energy contained in ocean waves. These waves are generated by the movement of winds across the world's oceans. The best wave regimes are usually found on a coast exposed to prevailing wind flow across a great ocean. The energy of the waves is contained in the oscillating motion of the water column formed by the sea at any point relative to the sea bed. A variety of devices have been designed to capture this energy, some based on floating buoys, others on floats that flap as a wave passes and yet others that isolate an oscillating column of water generated by waves reaching the shore to pump air through an air turbine. Designing devices of this type that are both robust and reliable has proved difficult.

OTEC is a technology that tries to exploit the temperature gradient found between the surface water in a tropical ocean and colder water a kilometre or more below the surface to drive a heat engine. To do this, cold water must be raised from the depths and used to condense water vapour that is created from the hot surface water by expanding it into a low-pressure chamber. A flow of low-pressure steam can be produced in this way and used to drive a turbine. An OTEC power plant requires a temperature gradient of at least 20_C to be able to operate effectively. Even so, efficiency is very low at around 6%.

Of the three marine technologies highlighted here, marine current generators are seen as the most promising. From the perspective of the energy resource, wave power offers the greatest potential if it can be mastered. OTEC has been demonstrated, but rarely, and commercial operation seems distant.

Power from waste is the term used to describe plants that can generate electricity from waste material of which there are presently copious quantities around the world. The most important of these plants are designed to burn municipal and some industrial waste. Most of this waste is collected within the world's cities, and in advanced economies it is — in theory at least — sorted so that any recyclable material is extracted. The residue can then be burnt in a special facility designed to completely destroy the waste and turn it into ash without releasing any noxious chemicals into the atmosphere. The heat from combustion, meanwhile, can be used to produce electricity. Such plants are very expensive to build and can only be operated economically if they charge a fee to dispose of the waste material delivered to them. Power from waste plants is therefore part of the waste disposal systems that operate in our cities and towns and electricity is a valuable by-product.

In the past — and in some regions still today — large volumes of waste have been buried in landfill waste sites. This waste, as it is covered and compacted, will over time start to generate methane gas as a result of the activity of natural microbes in the soil. This methane is then released from the surface of the landfill site. Methane is a potent greenhouse gas so the release is a significant environmental issue. However, if the gas is captured from within the landfill, it can be used to fire a gas engine and generate electric power. This particular type of waste to electricity is sometimes classified as renewable.

Fuel cells are electrochemical devices, closely related to batteries, that can convert a combustible fuel into electricity without actual combustion taking place. Most fuel cells are designed to use hydrogen as the fuel from which electricity is derived. The gas is fed to one electrode of a fuel cell while oxygen or air is delivered to the second electrode. With special catalysts, the reaction in which hydrogen burns in oxygen at high temperature can be carried out at relatively low temperature and much of the heat energy that would be released during combustion is converted, electrochemically, into electrical power. The fuel cell process is completely different to the heat energy cycle typically used for power generation from combustion fuel and this allows relatively high efficiencies to be achieved.

Hydrogen is not widely available and so for the most part fuel cells available today use natural gas. This is converted into hydrogen using an energy intensive process called reforming, which reduces the overall efficiency of electricity production.

There are four principles types of fuel cell that have been developed for power generation, the proton exchange membrane fuel cell (PEMFC), the phosphoric acid fuel cell (PAFC), the molten carbonate fuel cell (MCFC) and the solid oxide fuel cell (SOFC). The first two of these, the PEMFC and the PAFC, operate at relatively low temperature and require an additional reformer module to produce hydrogen from natural gas. The other two, the MCFC and the SOFC, operate at elevated temperatures and the reforming of natural gas can be carried out within the fuel cells or at their electrodes.

With the exception of the MCFC, all fuel cells are easily scalable and will operate with similar efficiency in 1 kW modules or as 1 MW power plants. They are relatively quiet and the only exhaust gases are water vapour and carbon dioxide. This makes them attractive for use in urban environments and as distributed generation modules. Some types have also been developed as domestic combined heat and power units.

Environmentally, fuel cells are considered benign. When operating with natural gas, they will generate carbon dioxide. It is possible to design fuel cells in which this gas is isolated and captured. However, the most interesting application for fuel cells is in a potential future hydrogen economy when hydrogen gas becomes the main combustion fuel in place of natural gas and gasoline. When burning hydrogen directly, fuel cells are extremely efficient and clean and could provide a range of benefits. Of especial interest is their use as transportation power units supplying electricity for an electric vehicle. Today, they are still relatively expensive but costs are falling.

Energy storage

The ephemeral nature of electric power makes it impossible to conserve electricity once it has been produced. This has always been a problem for the electricity supply industry because it means that supply must be available to meet demand and vice versa. There are ways of creating stores of electrical energy, but this invariably involves converting it into some other form of energy from which it can be recovered at will. Energy storage has been used to a limited extent in the past but energy storage systems are seen as increasingly important now that grids are having to accept large volumes of renewable power from unpredictable or unreliable sources. Under these conditions, energy storage provides the most convenient way of managing grids by putting excess power, from wind farms for example, away when it is not needed and then releasing it again when demand rises and the wind drops.

Hydropower power stations with reservoirs can provide a form of energy storage that can be exploited in this way but they are not always available because the water in the reservoir can become exhausted. There is a type of hydropower station, called a pumped storage hydropower plant, that is designed specifically for storage. This type of facility has two reservoirs, an upper and a lower. Water is stored in the upper reservoir until energy is needed, when it is used to drive turbines, and then captured in the lower reservoir. Meanwhile during periods of excess power on the grid, the surplus is used to pump water from the lower reservoir back into the higher. Plants of this type are generally large, up to several thousand megawatts in generating capacity, and are costly to build but very effective.

Rechargeable batteries, which store electrical energy in chemical form, are the most widely known form of electricity storage. They have traditionally been used for small-scale electricity stores, but over the past two decades they have been of increasing interest in using them on a much larger scale for grid energy storage. A variety of battery types have been developed for large-scale use, but the most common is the lead acid battery, common from use as starter cells for automotive vehicles. Lithium batteries are also beginning to be introduced. These are of particular interest for small-scale renewable energy storage at a domestic or small commercial scale where they can store the power from rooftop solar panels for later use.

There is a subset of this group of energy storage devices called flow batteries. A conventional rechargeable battery contains within its casing all the materials needed to store or release electrical energy so its capacity is fixed. A flow battery has electrodes like a standard cell but the chemicals that are used to either produce or store electricity are kept in external tanks. The amount can therefore be increased at will and their capacity is, in principle, unlimited.

However, development of cost-effective forms of this type of cell has so far proved elusive.

Compressed air energy storage (CAES) is a storage technology that, like pumped storage hydropower, has the potential to provide very large-scale electricity storage. The technology is based on that of the gas (or air) turbine that uses high-pressure air to drive a turbine generator and produce electricity. A conventional gas turbine has a compressor that provides high-pressure air, a combustion chamber where fuel is burned in this air to heat it and increase the pressure further and then an air turbine that uses this hot, high-pressure air to drive the generator. A CAES plant divides this scheme into two parts. When surplus energy is available, air is compressed using a compressor and then the compressed air is stored in a tank or underground cavern. When electricity is needed, this high-pressure air is extracted again and used to drive an air turbine. A combustor may be added at this stage to increase the energy available. There are a handful of plants using this technology in operation today but it has not been widely exploited.

Another way, potentially, of providing large quantities of stored energy for electricity production is via hydrogen. The gas can be produced from electricity by electrolysis of water. It must then be compressed and stored. The gas can be moved through pipelines like natural gas and can be used to generate electricity in most types of combustion power plant, in addition to being the ideal fuel for use in fuel cell power plants. However the production of hydrogen from water is relatively inefficient and to be cost-effective requires a supply of very cheap electricity. This might be supplied in the future from low-cost renewable energy such as that produced by large offshore wind farms.

In addition to these large-scale energy storage technologies, there are a range of systems that are intended for small-scale, often fast-acting, energy storage. These include energy storage capacitors, often called supercapacitors, flywheels and superconducting energy storage devices. Capacitors store electrical energy in the form of static electric charge across the plates of an electrical capacitor, but in a superconductor, this is supplemented electrochemically, allowing a single capacitor to carry a much higher charge. Once charged, the device can release electrical energy very quickly, making these devices ideal for high-speed grid backup systems.

Flywheels store energy in the rotational motion of a mass spinning at very high speed. As with a supercapacitor, this energy can be released very quickly and these devices can be used for high-speed grid support too. Some have also been used for much larger scale energy supply support such as for metropolitan railway transit systems.

Superconducting magnetic energy storage (SMES) devices are novel storage systems that rely on the ability of some special materials to reach a state of zero electrical resistance below a certain (often very low) temperature. As the electrical resistance falls to zero, materials of this type can carry a circulating electrical current without any resistive losses, so the energy contained in the electrical current can be maintained, in principle, indefinitely. In practice, this is not quite true but losses are low. There is a cost, though, because it is expensive to maintain the device at a very low temperature indefinitely. Small SMES systems are extremely fast-acting and have been used for grid support services. Larger scale SMES systems have been proposed in the past but they do not appear to be cost-effective today.

The use of large-scale energy storage for modern grid support and stability is extremely attractive but large systems are expensive and the structure of modern electricity industries can make it difficult to fund their construction. If electricity production is to be truly carbon free before the end of this century, then such energy storage is likely to become essential.

Chapter 3

Electricity networks

The production cost associated with the electricity-generating technologies outlined in Chapter 2 accounts for a large part of the cost of the production of electricity. This production cost is the cost of electric power as it leaves a power station and enters the electricity network. At this point, a number of other factors come into play and these may have a significant bearing on the actual cost of this electricity within the network and for the consumer. One of these factors is the structure of the electricity system, especially if it has been adapted to allow market forces to operate. Where such conditions have been put in place, the price may depend upon availability as well as production cost. If electricity is scarce, prices within the network will rise, and if there is a surplus, they will fall even though the end price may then have little relation to the actual cost of production. Not all electricity supply systems function in this way but many do. The alternative, a government owned and operated (or privately owned - strictly regulated) vertically integrated electricity supply system without any market element is rare now.

The structure of an electricity system does not lend itself naturally to market-based operation, and in many cases this is a recent development, its inception often driven ideologically. Some parts such as the electricity transmission and distribution networks are natural monopolies. Other parts such as power generation can be more easily turned into markets, but in doing so, governments lose the power to direct the way the industry develops and this can lead to strategic security-of-supply weaknesses. Separation of ownership, necessary for a market to work freely, can also lead to unexpected results. Nevertheless, an electricity supply system structured to accommodate free market principles is the norm today.

Historically, the electricity supply system in most nations grew in a piecemeal fashion with small, independent, privately owned networks supplying electric power from small power stations to a limited number of customers. As the number of networks grew and individual networks became larger, it became clear that greater efficiency could be achieved by aggregating the individual operators and their operations into national systems. By the middle part of the 20th century, the governments of many countries operated national electricity systems with similar contours.

The Cost of Electricity. https://doi.org/10.1016/B978-0-12-823855-4.00003-6

The traditional grid structure

The typical national electricity grid as it would have appeared in the 1960s or 1970s would have had a hierarchical structure[1] based around large central power stations. These power stations would have been connected to a national transmission system, a backbone network that could carry large volumes of electrical energy to all parts of the country. The transmission systems operated using high-voltage alternating current to reduce transmission losses.

In exceptional cases, such as for a very large industrial facility, power might be delivered from the transmission system directly to an end user. More normally, the power from the transmission system was fed into local distribution systems through transformer-based substations. These distribution grids covered a much smaller area and operated at lower voltage. Industrial and commercial consumers would receive power directly from this distribution system. Meanwhile smaller commercial consumers together with domestic consumers would be supplied through a yet lower voltage distribution network from which power cables could be taken to individual dwellings. Thus, power was produced centrally and delivered through a hierarchical delivery system so that power flows were always from the centre to the periphery.

At the centre of this web, there would sit a national control centre where the supply and demand on the grid was monitored and power stations were brought into service or taken off line as required. It was here that the various types of power plant duty cycle were defined. Base load power plants delivered constant output at all times into the grid: these were the cheapest to operate. Intermediate load plants would come on line in the morning and run through the day before backing off as load fell in the evening. And on top of these were the peak load (peaking) units that could be brought into service quickly as demand spiked. These fast-acting peaking units were the most expensive to operate.

A centrally owned and operated electricity system of this type can be developed in line with government strategy. The types of power plant that are built, when and where each is built and how capacity and network expansion is managed can all be decided at a national level based on a national security-of-supply considerations. This also allows costs to be controlled. The funding for new projects would usually be government backed, and large expensive projects such as a major energy storage plant could be built relatively easily.

Deregulation and grid privatisation

In the middle of the 1980s, a politically driven desire to eliminate these national electricity monopolies began to gain momentum in some countries such as the

---

1. In the United States, there was more than one grid, but most smaller nations operated a single grid.

United Kingdom and the United States[2] and soon spread elsewhere. The government-owned national systems were seen as inefficient and their practices antiquated. They were also becoming costly for the governments to underwrite. The solution was to break them up and sell the parts to private sector companies. To achieve this, the national monopolies were first split into separate transmission, distribution and generating companies. The transmission system was a natural monopoly necessitating a single, regulated company to be formed to own and manage it. The array of distribution networks were similarly divided into regulated regional monopolies that were sold off individually. Finally, the numerous power plants that together make the national generating company were sold, usually individually, to private operators.

In order to make this into a functioning marketplace, it was necessary to add a further layer, an electricity market into which power plants offered power for sale and consumers — these being either large industrial and commercial organisations or the regional distribution companies — that would purchase power. This market would support various long-term contracts between producers and consumers as well as a spot market for the instantaneous purchase of electrical power.

This 'deregulation' of electricity supply markets led to some notorious episodes including market manipulation and massive spot price spikes, but it was eventually tamed into the electricity supply system found in many parts of the world today.

One of the key features of a market of this type is that the players, now private companies seeking to make a profit, are usually driven by strict economic considerations when making strategic decisions such as the type of power plant to build. So, for example, there is little incentive for any company, be it a generating company or a distribution company to build an expensive energy storage plant because it will not provide the return that the private company investors seek. From a strategic national perspective, these decisions can often appear shortsighted, but a government can only try to encourage companies to behave strategically by introducing additional incentives.

Whether this way of operating electricity supply systems is the most effective method available is a matter for political debate and there are voices that question its utility. For the moment, however, it is the standard for most democracies and many less democratic nations.

Electricity prices, electricity supply contracts and the spot market

The electricity market in a deregulated electricity system is linked closely to the activities of the system operator which is charged with balancing the

---

2. In the United States, many of the large utilities were investor-owned, government-regulated, vertically integrated businesses. The same principles of deregulation were applied here too.

system so that supply and demand match. In a market system, the cost of the commodity in question will be determined by demand. The higher the demand at a particular time, the higher the price is likely to be. In the electricity marketplace, as in other similar energy markets, the varying hourly and daily cost of electricity is called the spot price.

The system market operator which manages the functioning of the market will police that sale and purchase of electricity with the spot price signalling the value of electricity at any given time. A rising spot price is an indication that demand is high and increasing and this will encourage generators to bring into service as much capacity as they can. Conversely, when the spot price is falling, generators will back out of the market with their most expensive generating units. On the other side, consumers such as distribution companies will have to pay the spot price to be able to maintain supply to their clients. However, this can leave both sides vulnerable to wild swings in costs. For example, while the average spot price in different regions on the Australian National Electricity Market in the year to June 2018 was between Aus$73/MWh and Aus$98/MWh, the price could theoretically rise to $14,500/MWh before it is capped.[3]

To combat this danger, consumers can hedge their costs by striking contracts for electricity with generators. These contracts ensure that even if the spot price for electricity should rise dramatically, the consumer will only pay to price agreed by the contract for the amount of electricity covered by that contract. Of course if the contract does not cover all the needs of the consumer, then it will still have to go to the spot market for the remainder and pay the spot price. The contracts, which are always for future supply when made, will be at a price that reflects the expected spot price for power when the electricity must be supplied. Contract of this type allows both generators and consumers to make economic plans with the security of knowing that these contract prices will be stable. If a power plant is contracted to supply power to a distribution company at a particular fixed price and it does not do so, then it will be penalised at a level relating to the spot price at the time when it reneged.

There is another type of power supply contract called a Power Purchase Agreement (PPA) that can be struck between a power generating company and another party — usually a distribution company, but in some instances an industrial facility — in which the buyer contracts to buy all (usually) of the output from the power generating facility over the term of the contract, which is usually for a long period and at a fixed price. This type of contract is often used by renewable generating companies seeking to provide the financial security to be able to develop the project. It will offer long-term security but with the caveat that the cost agreed when the contract is struck may be wildly different from the actual market price of electricity over 10, 20 or 30 years.

---

3. Australian Energy Market Commission.

Wholesale and retail cost of electricity

While distribution companies purchase power using contracts and the spot market, they offer this power for sale to their customers at a fixed price, the tariff, which is determined by the contract between the two parties. The tariff price may vary with the time of day, and from time to time, the prices will change as costs to the distribution company change, but consumers will normally be able to predict the cost of their energy at least over a horizon of a few months.

This consumer tariff will be significantly higher than the wholesale price that the distribution companies pay for their electricity. For example, in Australia, the wholesale price paid by the distribution company typically accounts for only 30%–40% of the consumer price.[4] Another 40%–50% will be accounted for by the transmission network costs (including profits for the operator). There is a 5%–15% addition to cover the environmental costs resulting from government schemes and then a further 5%–15% that covers the residual costs of the distribution company and provides the profit that this company expects to make on its sales.

The difference between the wholesale and the retail cost will vary from country to country, but a more than doubling of the end cost to the consumer is probably typical. These end costs will also vary from consumer to consumer. For example, in the New England region of the US grid in the middle of 2018, residential customers paid just over $200/MWh, commercial customers paid around $160/MWh while industrial customers paid perhaps $120/MWh.[5]

These price variations can have important implications for some types of power generation. A household that is considering installing solar cells onto the roof in order to generate electric power will break even if it can undercut the retail cost of electricity from the grid. This, in turn, means that manufacturers of rooftop solar installations for domestic use do not need to be able to provide technology that can undercut the wholesale price when the local retail price is much higher. This can make even relatively expensive renewable installations competitive in the right situation.

The later chapters of this book will be primarily concerned with the actual cost of production of electricity from different types of generator. This will be most closely related to the wholesale cost of production of electrical power before it actually enters the network.

Distributed generation

The introduction of an electricity market into an electricity supply system does not by itself affect the overall structure of an electricity system which remains

---

4. Australian Energy Market Commission.
5. US Energy Information Administration https://www.eia.gov/todayinenergy/detail.php?id=37415.

hierarchical after these changes have been introduced. However, the introduction of market forces coupled with the changes in the way electricity is being generated has begun to break down this strict hierarchical structure. For example, consumers large and small may seek to generate their own electricity either for local consumption or to sell back to the market. At the same time, a variety of small-scale generation technologies have grown up, technologies that make local generation more easily achievable. This is the purview of what is known as distributed generation.

Distributed generation refers to the production of electric power at the periphery of the grid rather than at the centre, so that the source of electricity is much closer to the consumer. Rooftop solar panels are a perfect example of this type of generation. When a domestic dwelling installs solar cells on the roof, these generating devices provide power directly to the household, not across the grid, so any grid losses are eliminated. In many cases, the electricity from these units is used exclusively by the household, perhaps with local domestic storage batteries to manage any excess, unused power. This power production and consumption all takes place on the consumer's side of the electricity meter.

There are increasing instances, however, where electricity generation at the household level is not restricted to that household circuitry alone. Often there are arrangements with the local distribution company which allow surplus electricity to be exported across the domestic meter and sold to the distribution company, thereby entering the distribution grid. And there may even be laws that require that distribution companies purchase this electricity.

In addition to this type of household production, small power stations are being designed to supply power to the distribution grid rather than to the transmission system. The advantage of this is, again, that the electricity is produced much closer to the consumers and that means that energy losses from the transmission and distribution of the electricity are much lower. So while the cost of electricity from a small power station might be higher than from a large central power station, the proximity to the consumer makes the price lower than it would be if the same power was supplied from the central power station.

A further advantage of this type of generation is that it allows generating capacity to be increased in much smaller tranches as demand rises. Instead of needing to fund a massive central power station, a generating company may choose to build 5 or 10 small stations as required.

Once this type of generation is embraced, there are any number of different configurations that can be envisaged. A local community might decide to build its own power station using wind turbines and solar cells, selling the power to the members of the community and delivering it across the local distribution network. The distribution company will charge for the service, but it may still allow the community power to undercut the grid supply. Or, a company that has installed backup generators in case of power supply failure might chose to run these units at times of peak demand and sell the power they produce to the grid.

Distributed generation has many attractions both in terms of cost and energy efficiency. But, the introduction of distributed generation means that the distribution network is no longer a passive network that takes power from the transmission system and delivers it to the consumer. Now there are power flows into the distribution network from both the transmission grid and the generators that are connected directly to the distribution grid. In extreme cases, there may be power flows from the distribution network back into the transmission network. The distribution network has thus become an active rather than a passive network, and it needs its own network control centre to manage the energy that is running through it. Life for distribution companies has suddenly become much more complicated.

The smart grid

For distribution network operators, the relatively simple process of buying power from the national market and then selling it to a consumer has become more of a matter of balancing the amount of power on the distribution network, from both the traditional suppliers and the new distributed generators, against demand. Meanwhile at the transmission grid level, there are new complications resulting from both the potential flow of power from distribution networks into the transmission network and the variable supply from wind and solar power plants, all of which must be balanced against demand. These are some of the factors that have led to the development of what has become to be known as the smart grid.

The smart grid is, in essence, the addition of a computer communications network that can run alongside the power distribution network, mimicking it. Once this network is in place, intelligent devices — computers — can be installed at key points in the network and these devices can communicate with one another and so help manage the electricity network itself. This may be as simple as an intelligent meter (smart meter) in a household that can signal back to the network control centre the amount of power that household is using and what machines are using it. In times of high demand, the control centre operator may then be able to ask some of the machines in the household to shut down until demand drops. This is known as demand management.

Another smart grid technology is congestion control. Power networks are physical structures, much like roads, and if too much power is trying to pass down the same power line, congestion occurs. Sensors placed at key points across the network can monitor these flows, just as traffic can be monitored and controlled using close circuit TV cameras, and the information from these sensors can be used to control how much power passes down each line. At the limit, automated devices can act autonomously to maintain the stability of the grid. This type of technology is a key to managing a distribution network that must now be capable of operating pro-actively.

Smart grid techniques are equally important for the transmission system. One of the most important areas in which these technologies can help with grid management is with the integration of renewable energy into the supply. Some of these energy sources, but particularly wind and solar power, are both variable and unpredictable. Balancing the grid with these uncertain sources can be tricky, particularly as they often provide the cheapest power when they are generating and so this must be dispatched first. One of the smart grid tools to help manage renewable power is weather forecasting. If accurate day-ahead or, better, hour-ahead forecasting is available for features like wind speed over different parts of the region and cloud cover in the case of solar generators, this can automatically be fed into the grid management control centre and changes in renewable output can be predicted and countered ahead of time.

Another valuable way in which smart technology of this type can help with renewable energy is via the concept of a virtual power station. Wind power provides a good example of how this concept can be advantageous. Wind speed varies from hour to hour and day to day, and at any single wind farm site this will lead to a wildly varying output. However, the variation in wind intensity over a wide geographical region can be much lower. If the wind is not blowing in one place, it will probably be blowing in another. If several wind farms, separated geographically, are combined into a single, virtual power station, the output from the power station will be much less variable and therefore much more reliable than from any one of them alone. In effect the wind is being averaged, geographically. This makes the management of the output from the wind plants easier and the electricity they produce more valuable.

This concept can be expanded to include other generating technologies, a technique called power pooling. Wind, solar power and even some fossil fuel generation might be included in a single virtual power plant if the result was advantageous from an economic and reliability perspective.

Microgrids

One of the new concepts that emerged from the combination of distributed generation and the smart grid is the microgrid. A microgrid is a small, self-sufficient grid that includes a defined set of consumers together with power generation facilities that are capable of supplying them with all the power they need. While these technologies might include small fossil fuel generation units such as gas engines, many will be designed to use renewable power from wind and solar sources. And because these cannot provide a reliable supply alone, a microgrid of this type will also need some form of energy storage.

A microgrid is not isolated; it is connected to a part of a larger distribution network and it can both take power from that network and deliver power back into it. However, it is managed locally as a single unit, and power delivery across the microgrid is managed using local smart grid technologies. Most importantly, the microgrid is capable of operating as an isolated grid if the main distribution grid should fail. This resilience is a key part of the concept.

It has been suggested that microgrids might form the building blocks of future national grid systems. Taken to this extreme, a national grid would simply be an array of interconnected microgrids that use smart technologies to cooperate as a single large grid but in the event of failures can fall back on themselves. This idea presupposes that all electricity generation takes place at the microgrid level, something that seems extremely unlikely given the economic advantages of larger power plants. Even so the concept could be important as an element of future hybrid grid structures.

Electric vehicles

One of the most important sources of greenhouse gases, aside from fossil fuel power plants, is that generated by transportation, by vehicles that use petrol and diesel fuel as their energy source. One of the solutions to that problem today is the electric vehicle that uses an electric motor to provide motive power. That motor is in turn powered by a rechargeable battery, and in order for this technological solution to be effective, vehicle batteries must be recharged regularly.

Electric vehicles present the electricity industry with a major problem. A wholesale shift to electric vehicle power will lead to an enormous increase in the load on the grid. This new load must be managed and smart grid technologies are likely to be essential to enable this to be carried out smoothly.

At the same time, electric vehicles potentially offer a solution to one of the great problems of the modern grid, a lack of energy storage. A massive population of electric vehicles can also be seen as a massive collection of energy storage devices, the vehicles' batteries. If the aggregate capacity of these batteries could be harnessed, then it could help manage the integration of renewable energy into the grid and reduce overall grid operating costs, without the need for large energy storage plants.

Achieving this in practice will be difficult. It is only when an electric vehicle is connected into the grid that its storage capacity can be harnessed for grid support purposes. While it is likely that a large number of vehicles will be connected and charging at any one time, there needs to be a way of managing their number: a sudden rush to the seaside by half the population on a hot day might bring the grid down, otherwise. Nevertheless, the concept of distributed storage capacity based on vehicle batteries provides an interesting and potentially extremely valuable extension of smart grid technologies.

Supergrids

At the other end of the scale from the microgrid in terms of size and ambition is the supergrid. At its simplest, a supergrid is a supranational (or supraregional) grid that acts to combine the grids of different nations so that they can share their electricity generation resources. A scheme of this sort has been proposed to integrate Europe's national grids and allow power to be moved from country to country more easily than is possible now. There are also some special supergrids that have been proposed to solve particular energy harvesting and supply issues.

Supergrids are based on high-voltage direct current (HVDC) power lines that can move power over long distances with lower losses than would be found with the equivalent alternating current power line. Elements of supergrids already exist within national certain grids. In the United States, for example, there are some HVDC interties that connect regional grid systems together, while in China, HVDC power lines are used to move power from regions with large generating capacities to centres of demand. However, neither of these applications can be viewed as a coherent supergrid.

Some ambitious supergrid schemes have been proposed in the last decade or two, though none has yet been constructed. One such is an offshore supergrid in the seas of western Europe. Such a grid might run from Scandinavia all the way to Portugal. The purpose of such a grid would be to bring offshore wind power from these waters onshore and to deliver it to the nations along the western coast of Europe. However it would also provide a way of interconnecting grids of the various seaboard nations so that they can exchange power too. A subset of this is a proposal to build a North Sea supergrid to serve wind farms in the North Sea and at the same time to interconnect the nations with North Sea coasts.

A second supergrid scheme that has been proposed would link grids in southern Europe to those in North Africa. The latter has massive amounts of solar energy available, which could be captured in large solar power plants built in the desert regions of the countries of North Africa. This power could then be exported to the demand centres in Europe through the proposed supergrid.

Much of the work on supergrids is speculative. However, the ideas being developed for this, and other smart grid concepts, are likely to find their way into the electricity systems of the future in one form or another.

Chapter 4

Fuel costs

The cost of electricity from a power station depends on a number of different factors, but the two most important are the capital cost of building the facility and the cost of the energy that it uses to produce electricity. For many types of power station, the source of energy that is used to make electricity is a combustion fuel such as coal, natural gas or oil. Others burn biomass-derived combustion fuels, and in the future, some stations will burn hydrogen. All these fuels are more or less transportable and the fuel will be delivered to the power station which will pay for each unit of fuel it receives.

Nuclear power stations also require fuel, in this case a specialised fuel that is normally based on uranium. This fuel, which is fabricated into special fuel rods, is also delivered to the power station where it is utilised and again there will be a cost attached to each unit.

There are other types of power station, renewable energy plants that utilise an energy source that is generally considered free to use. While it is possible in some cases to claim ownership of the resource that is being exploited, for example, ownership of the land upon which a hydropower plant is built, this is unusual. The energy from the sun, the wind and the energy in flowing water are normally available to turn into electricity without any cost attached to the consumption of the energy.

While the energy is free, the use of renewable energy will often impose other constraints on power station construction. Hydropower plants can only be built at river sites where the water flow is suitable for exploitation. This may mean building a power plant far from the main centres of consumption. Wind power also relies on a suitable resource being available. Some of the best wind sites are offshore, but these are also much more difficult to exploit than onshore wind sites. Solar power is somewhat more equitable in this regard, but the economics of utility scale solar power plants will restrict the available sites, particularly where solar thermal power generation is concerned.

The effect of these constraints will be to increase the capital cost of construction of the renewable power plant. If the facility is located far from the main grid, then additional cost will be attached to the construction of a power line to carry the power to the grid. The location may also make the job of construction more difficult. In contrast, a combustion plant can be built anywhere, and major coal and natural gas-fired power plants are usually located close to the transmission grid.

The Cost of Electricity. https://doi.org/10.1016/B978-0-12-823855-4.00004-8

These factors may affect the capital cost of construction of a power plant, but there is, nevertheless, a fundamental difference between power plants that require a combustion fuel and those that use a renewable energy source. For the former, there will always be a fuel cost associated with the production of electricity. For a renewable energy plant, in contrast, the cost of production will be very small once the capital cost of the power plant has been met. Hydropower plants, once their financial costs have been paid down, provide some of the cheapest power available. As these power plants can have extremely long operational lives, this situation is not uncommon.

Combustion fuels

The main combustion fuels for power generation are coal and natural gas. In the past, oil was an important source of electricity too, but most of the oil that is pumped is used elsewhere today. However, there is still limited use of oil in small piston engine power plants, and some large combustion power stations in oil-producing countries continue to burn the fuel.

Oil and its byproducts are valued because they are liquid fuels, they can be easily transported and a small quantity contains a large amount of energy. Crude oil typically contains 46 MJ/kg.[1] The high energy density and the portability are reflected in the cost of the fuel which is the most expensive of the fossil fuels. Natural gas can be delivered by pipeline and it can be distributed in liquefied form, but it is less adaptable than traditional liquid fuels. It has a higher energy density than oil, 54 MJ/kg, but its volume energy density is much lower. Coal, the most popular fuel for power generation, is the least easily transported of the major fossil fuels. Its energy density, typically 24 MJ/kg, is lower than either oil or natural gas and transportation costs can be high. In consequence, coal is often used close to the source, the mine. Some high-quality coal is transported over large distances but most is consumed locally.

Table 4.1 shows the variation in the cost of these fuels on a unit of energy basis when they are used for power generation in the United States. While local factors will affect the actual figures, the relative costs are generally similar in most parts of the world, but with a caveat concerning the cost of natural gas, noted below.

The most expensive of the fuels is high-quality distillate fuel, which, in the United States, costs $15.16/million BTU in 2019. Residual fuel oil, the lowest quality fuel oil, costs $12.72/million BTU. Natural gas for power generation had a typical cost of $2.88/million BTU, 19% of the cost of a similar amount

---

1. Power Generation Technologies, third edition, Paul Breeze, Newnes 2019.

TABLE 4.1 Average US fuel costs for power generation, 2019.

Fuel	Power generation fuel costs ($/million BTU[a])
Coal	2.02
Natural gas	2.88
Residual fuel oil	12.72
Distillate fuel oil	15.16

Source: EIA[b].
[a] *1 million BTU = 1.056 GJ.*
[b] *Short-Term Energy Outlook — June 2020, US Energy Information Adminstration.*

of distillate fuel, in energy terms. Meanwhile, coal, the cheapest fuel in Table 4.1, had a cost in 2019 of $2.02/million BTU or 13% of the cost of distillate fuel.

The cost difference between natural gas and coal in the table is relatively small. This, at least in part, reflects the low cost of natural gas in the United States in consequence of the expansion of shale gas production there. The difference in cost between coal and natural gas is likely to be much larger in other parts of the world. Given the prices differences shown in the table, it is no surprise that oil is rarely used for large-scale electricity production.

Table 4.2 compares the annual consumption of these fuels in the United States for power generation between 2010 and 2019. The most notable feature in the table is the decline in the use of coal for electricity generation over this period. Consumption of coal in 2019 at 7,919,000 billion BTU was only 56% of the level in 2010. Over the same period, the annual consumption of natural gas grew, from 3,396,000 billion BTU to 5,436,000 billion BTU, an increase of 60%. These figures are indicative of a switch from coal to natural gas over this period in the United States. Similar trends will be found in many advanced nations as a result of environmental concerns. Elsewhere the picture is not quite so clear with coal consumption for power generation still increasing in China and India. The other observation to make from the table is that petroleum liquids have only limited use for power generation in the United States in the 21st century and that usage is declining. This reflects the fact that petroleum liquids are generally too expensive to use in power plants.

Oil

Crude oil is a liquid hydrocarbon formed from organic material buried and compressed within rock in the earth's surface. The composition of crude oil varies but mostly contains between 82% and 87% carbon and 12%−15%

TABLE 4.2 Annual fuel consumption for power generation in the United States (billion BTU).

Year	Coal	Petroleum liquids	Natural gas
2010	14,227,000	190,000	3,396,000
2011	13,872,000	144,000	3,571,000
2012	11,940,000	86,000	4,084,000
2013	11,595,000	78,000	3,939,000
2014	12,065,000	98,000	3,877,000
2015	11,089,000	90,000	4,718,000
2016	9,257,000	73,000	5,075,000
2017	9,012,000	70,000	4,794,000
2018	8,351,000	84,000	5,564,000
2019	7,919,000	65,000	5,436,000

Source: EIA[a].
[a]*Electric Power Monthly May 2020, US Energy Information Administration. The figures in the table are based on power plant fuel receipts.*

hydrogen by weight. The most important constituents of crude oil are paraffins, naphthalenes and aromatics. Paraffins are the main constituents of gasoline, used for transportation, while aromatics and naphthalenes are often used by the chemical industry as precursors for other products.

Global oil reserves are not distributed evenly across the globe. The largest regional reserves are found in the Middle East with 834 thousand million barrels of proven reserves at the end of 2019.[2] Largest reserves were in Saudi Arabia, Iran, Iraq and Kuwait. South America also has significant reserves, primarily as a result of those in Venezuela, which total 304 thousand million barrels. Other countries with notable reserves include Canada with 170 thousand million barrels, the Russian Federation with 107 thousand million barrels and the United States with 69 thousand million barrels. In Africa, only Libya, Nigeria and Algeria have significant reserves and the proven reserves are limited in the Asia Pacific region too.

When crude oil is extracted from the ground, it is usually sent to a refinery where it is separated into different constituents that can be used for a variety of purposes. The processing leaves a residual oil that is often used for power generation.

2. BP Statistical Review of World Energy 2020.

The standard measure of crude oil is the barrel, a unit defined by the US oil industry. A barrel of oil is equivalent to 42 US gallons of oil, or 159 L. Elsewhere in the world, weight is often used to define crude oil quantities. The relationship between volume and weight depends on the quality of the crude oil, but a barrel of light oil might weigh around 140 kg.[3] A tonne of a similar oil would contain just over seven barrels. Oil markets usually price oil per barrel.

Table 4.3 shows figures from the *BP Statistical Review of World Energy 2020* for the average annual price of a barrel of Brent Crude between 1989 and 2019, adjusted to $2019 to take account of inflation. As the table shows, the cost of a barrel of oil can be extremely volatile. The price often depends on global economic conditions, with a high price typical during periods of rapid economic growth and a low price when there is a slump. Geopolitical factors also play an important role. When oil supply is threatened in a particular part of the world, prices are likely to rise.

The average cost in 1989 was $37.58/barrel and the average prices (not shown) were relatively stable during the next decade, dropping to a low of $19.94/barrel in 1998. The price began to rise during the next decade, drifting upwards so that near the end of the decade, in 2008, they peaked at $115.48/barrel before falling back sharply the next year, to $73.49/barrel, as a result of the global financial crisis. This dip was short-lived, and by 2011, the average price was $126.45/barrel. This proved to be a peak. The cost dropped abruptly in 2015 to $56.51/barrel and remained in this region until 2019 when the average price was $64.21/barrel. However, the average price is likely to show a sharp fall again in 2020 as a result of the coronavirus pandemic which pushed down consumption during the early part of the year. This, coupled with a policy of oversupply from OPEC and Russia, led to the weekly price of Brent Crude dipping to below $25/barrel at one point in April 2020.[4]

There has, historically, been a strong correlation between the price of oil and the price of natural gas with a rise in the price of oil pulling up the price of natural gas and a fall in the oil prices depressing natural gas prices. This linkage appeared to have broken between 2008 and 2019, at least in the United States, and since 2008 the prices of the two commodities have not tracked one another so closely. This has been put down to the rise in production of shale gas in the United States, which has depressed natural gas prices. However the rupture of the link may be less strong in other regions where there is no — or limited — domestic natural gas production. Production of oil and natural gas are linked because in most cases they are taken from the same wells and so some connection remains inevitable. It is conceivable, too, that a stronger correlation between the two may return.

---

3. Figures are taken from Encyclopedia Britannica.
4. Weekly Brent, OPEC basket and WTI crude oil prices from 30 December 2019 to 22 June 2020, Statista.

TABLE 4.3 Average annual oil price 1989–2019.

Year	Cost (2019$/barrel)
1989	37.58
1999	27.58
2000	42.31
2001	35.29
2002	35.56
2003	40.06
2004	51.79
2005	71.37
2006	82.61
2007	89.26
2008	115.48
2009	73.49
2010	93.20
2011	126.45
2012	124.35
2013	119.25
2014	106.85
2015	56.51
2016	46.59
2017	56.52
2018	72.60
2019	64.21

Source: BP[a].
[a]*BP Statistical Review of World Energy 2020. The figures in the table are for Brent Crude.*

Natural gas

Natural gas, like crude oil, is found by drilling into underground rock strata. Like oil, it is the product organic material that has been buried within the surface of the earth over aeons. When it emerges from the ground, natural gas is a mixture of combustible hydrocarbons. The main component is methane, which normally accounts for 70%–90% of the total. Other hydrocarbons such as ethane, propane and butane can account for up to 20% and there may be up to 8% carbon dioxide, small amounts of oxygen and nitrogen and up to 5% hydrogen sulphide. The gas is normally cleaned after it has been pumped from the ground to remove impurities such as hydrogen sulphide (this can be processed into pure sulphur), carbon dioxide and water. The higher hydrocarbons such as propane and butane may also be removed for industrial use leaving the cleaned gas, now referred to as dry natural gas. This is the gas that is typically pumped into pipelines.

Natural gas is found in many parts of the world, but most countries have only small reserves. Regionally, the Middle East has the largest total proven recoverable reserves, 75.6 trillion cubic metres or 38% of the global total at the end of 2019.[5] Iran alone commands 16% of global reserves and Qatar a further 12%. Meanwhile the Russian Federation holds the largest single national reserve, 19% of the global total. Turkmenistan has a further 10% and the United States 7%. These five countries together command close to two-thirds of the world's proven natural gas reserves.

Unlike oil, there is no global benchmark for the price of natural gas and the cost varies from region to region. Table 4.4 contains figures for average annual costs in $/million BTU in the United States (Henry hub), the import price in Germany which is typical of the price in western Europe and the cost of liquefied natural gas (LNG) imported into Japan where there are no indigenous reserves that are exploited. The final column compares the cost of crude oil for the same quantity of energy. Looking across the table during the early years for which figures are presented suggests that the prices are broadly similar across the globe and that the gas price is not dissimilar to that of crude oil. For example, in 1999, the average cost of a unit of gas at Henry Hub was $2.27/million BTU while the cost of a unit of crude oil was $2.98/million BTU, while 2 years later the average prices of these two commodities were $4.07/million BTU and $4.08/million BTU. The cost of LNG in Japan was slightly higher in both years, a difference to be expected when the fuel must be shipped from gas-producing nations to Japan. Meanwhile, the cost of natural gas in Germany which is delivered by pipeline from the Russian Federation, from Norway or from North Africa was slightly lower than at Henry Hub in the United States.

This pattern of prices broadly held until 2007, although by this date the price of a unit of crude oil had begun to rise above that of natural gas in most of markets shown in the table. The regional natural gas prices in the table were still close to one another, although the cost of imported gas in Germany has by 2007 risen above both Japanese LNG and the price at Henry Hub in the United States.

---

5. BP Statistical Review of World Energy 2020.

TABLE 4.4 Average annual natural gas prices ($/million BTU).

Year	USA, Henry Hub	Germany, import	Japan, Liquefied natural gas	OECD, crude oil
1999	2.27	1.86	3.14	2.98
2000	4.23	2.91	4.72	4.83
2001	4.07	3.67	4.64	4.08
2002	3.33	3.21	4.27	4.17
2003	5.63	4.06	4.77	4.89
2004	5.85	4.30	5.18	6.27
2005	8.79	5.83	6.05	8.74
2006	6.76	7.87	7.14	10.66
2007	6.95	7.99	7.73	11.95
2008	8.85	11.60	12.55	16.76
2009	3.89	8.53	9.06	10.41
2010	4.39	8.03	10.91	13.47
2011	4.01	10.49	14.73	18.55
2012	2.76	10.93	16.75	18.82
2013	3.71	10.73	16.17	18.25
2014	4.35	9.11	16.33	16.80
2015	2.60	6.72	10.31	8.77
2016	2.46	4.93	6.94	7.04
2017	2.96	5.62	8.10	8.97
2018	3.13	6.62	10.05	11.68
2019	2.53	5.25	9.94	10.82

Source: BP[a].
[a]BP Statistical Review of World Energy 2020. The figures in the table are for Brent Crude.

By 2008, there were signs of a change in the relative prices across the table and this was amplified in the succeeding years. The most immediately noticeable feature is that the price of a unit of gas at Henry Hub in the United States had fallen dramatically compared to both the cost of gas in Germany and Japan and the price of a unit of crude oil. In 2012, for example, the cost of a unit of gas in the United States was $2.76/million BTU while the cost in Germany was $10.93/million BTU and in Japan the average LNG price was $16.75/million BTU. The crude oil price was $18.82/million BTU.

The cause of this change was the sudden availability of large quantities of shale gas in the United States, which served to depress the price there compared to other parts of the world. It also severed to linkage between natural gas and oil in the United States because the rise and fall in natural gas prices no longer tracked that of oil.

In contrast, the prices for natural gas in Germany and for LNG in Japan continued to show some connection to the oil price. There was quite close correlation between Japanese LNG and global oil prices and both continued to track the global economic situation with a slump after 2008 as a result of the global financial crisis followed by a rapid recovery between 2010 and 2014. Prices of both then dipped during 2015—17 before climbing again in 2018. The correlation between oil price and the cost of imported gas in Germany was less clear after 2008 with the price differential increasing markedly after that year. Natural gas prices in Germany during this period reflected the changing market for natural gas in Europe. So while there was a clear global linkage between oil and natural gas prices at the beginning of the century, by the end of the second decade of the 21st century, natural gas prices showed a strong local component.

Natural gas is a much cleaner fuel than coal, so switching generation from coal to natural gas can reduce greenhouse gas emissions significantly. In addition, natural gas—fired power plants are relatively cheap to build. In the United States, a decade of relatively low natural gas prices has encouraged utilities to burn natural gas in preference to coal, as was shown in Table 4.2. The experience has not been quite so simple in other parts of the world. Some utilities in Europe also chose to build natural gas—fired power plants while gas prices were low, but the volatility that still exists in the natural gas price in most parts of the world has led to these plants becoming uneconomical to operate when prices rise too far. The situation has been aggravated in Western Europe by a glut of cheap coal from the United States as a result of coal being displaced there by natural gas. This has made generation from coal relatively more economical during the middle of the second decade of the 21st century.

It is important for power plant developers to be aware of the risk that can be associated with fuel price volatility, particularly with natural gas. It is never safe to assume that prices will remain low. There is an additional problem too. In the past if natural gas prices rose, utilities could increase the price of their electricity to compensate. Today the large generating capacity based on renewable resources available in many countries sets a backstop or hedge price. The renewable resources may be unreliable over the short term, but over the long term, they provide a stable supply at a stable price. Natural gas—fired plants must be able to undercut this price if they are to remain competitive. What is more, the stable, long-term cost of renewable energy is falling.

Coal

The term coal embraces a range of materials. Within this range, there are a number of distinct types of coal, each with different physical properties. These properties affect the suitability of the coal for power generation.

The hardest coal is anthracite. This coal contains the highest percentage of carbon (up to 92%) and very little volatile matter or moisture. When burnt, it produces little ash and relatively low levels of pollution other than carbon dioxide. Anthracite is typically slow-burning and often difficult to fire in a power station boiler and it has traditionally been used for heating rather than industrial use. However, it is becoming more common as a power plant fuel in countries with large reserves such as Russia, which have switched to anthracite for national power generation in order to free natural gas for export.

The most abundant of the coals are the bituminous coals. These coals contain significant amounts of volatile matter. When they are heated, they form a sticky mass, from which their name is derived. Bituminous coals normally contain between 76% and 86% carbon (dry content). Moisture content when mined is between 8% and 18%. They burn easily, especially when grounded or pulverised. This makes them ideal fuels for power stations. Bituminous coals are further characterised, depending on the amount of volatile matter they contain, as high, medium or low volatility bituminous coals. Some bituminous coals contain high levels of sulphur which can be a handicap for power generation purposes.

A third category called sub-bituminous coals or soft coals are black or black-brown. These coals contain between 70% and 76% carbon (dry content) and 18%−38% water as mined, even though they appear dry. They burn well, making them suitable power plant fuels, and sulphur content is low.

The last group of coals that are widely used in power stations is lignite. These are brown rather than black and have a carbon content of 65%−70% after they have been dried but the moisture content is 53%−63% when taken from the ground. Lignites are formed from plants which were rich in resins and contain a significant amount of volatile material. The amount of water in mined lignite, and its consequent low carbon content, makes the fuel uneconomic to transport over any great distance. Lignite-fired power stations are generally found adjacent to the source of fuel.

A type of unconsolidated lignite, usually found close to the surface of the earth where it can be strip-mined, is sometimes called brown coal. (This name is common in Germany.) Brown coal has a moisture content of around 45%. Peat, which has a dry carbon content of less than 60%, is also burned in power plants, though rarely.

Coal is found in most parts of the world. The largest regional reserves are in the Asia Pacific region where China, India, Australia and Indonesia all have significant amounts of all types. The United States also has very large reserves of coal. In Europe, Ukraine and Poland have large amounts of anthracite and bituminous coals, while Germany and Turkey have sub-bituminous coals and lignite in abundance. There are limited reserves in South and Central America, the Middle East and in Africa, where only South Africa uses coal extensively.

The value, and therefore the cost, of these different coals vary enormously. In the United States, in 2018, the cost of a tonne of anthracite was $110, for bituminous coal the cost was $66/tonne, the average price paid for sub-bituminous coal was $15/tonne while for lignite it was $22/tonne.

The average cost of coal in different regions of the world between 1999 and 2019 is shown in Table 4.5, based on figures from the *BP Statistical Review of World Energy 2020*. Traditionally, coal prices have not been as volatile as those of oil or natural gas, but as the figures in the table show, there was a significant increase in the cost of a tonne of coal in all regions in 2008 as global growth surged, just before the global financial crisis.

As coal is mostly consumed where it is mined, coal prices should reflect as strong local dimension. Nevertheless, the variations between different regions listed in the table are not large.

In the United States, the average cost of a tonne in 1999 was $31.29. The price moved both up and down during the next 8 years before peaking at $117.42/tonne in 2008. Prices quickly fell back again to $60.73/tonne in 2009, after which the price remained in a band between $84.75/tonne and $51.45/tonne.

The cost of coal in Northwest Europe was broadly in similar territory over the 20 years shown in the table but with somewhat greater volatility than in the United States and with a peak of $147.67/tonne in 2008 and another of $121.48/tonne in 2011. Japan imports all its coal. Even so the prices in Table 4.5 show the cost of coal in Japan to be competitive with that in Europe and the United States, at least until 2008. However, the spike in prices in Japan in 2008 was not tempered until 2014, with the price reaching a peak of $136.21/tonne in 2011. Prices began to fall back slightly after 2013 but rose again from 2017.

The spot price for coal in China is shown in the final column of Table 4.5. Here prices were relatively stable during the first years of the 21st century, rising slowly towards the end of the first decade when they peaked at $104.97/tonne in 2008. There was a further peak in 2011, after which prices stabilised, but at a much higher price than in the first decade of the century. In 2019, the average price was $85.89/tonne, higher than in either the United States or Northwest Europe.

Biomass

The term biomass fuel encompasses a wide range of materials that can be used for power generation, as noted in Chapter 2. Waste materials, particularly agricultural wastes, are one of the important local sources and where they are available they can be used as combustion fuel in small combustion power

TABLE 4.5 Average annual coal prices ($/tonne).

Year	USA, Central Appalacian spot price	Northwest Europe	Japan, Steam coal	China, spot price
1999	31.29	28.79	35.74	n/a
2000	29.90	35.99	34.58	27.52
2001	50.15	39.03	37.96	31.78
2002	33.20	31.65	36.90	33.19
2003	38.52	43.60	34.74	31.74
2004	64.90	72.13	51.34	42.76
2005	70.12	60.54	62.91	51.34
2006	57.82	64.11	63.04	53.53
2007	49.73	88.79	69.86	61.23
2008	117.42	147.67	122.81	104.97
2009	60.73	70.39	110.11	87.86
2010	67.87	92.35	105.19	110.08
2011	84.75	121.48	136.21	127.27
2012	67.28	92.50	133.61	111.89
2013	69.72	81.69	111.16	95.42
2014	67.08	75.38	97.65	84.12
2015	51.57	56.79	79.47	67.53
2016	51.45	59.87	72.97	71.35
2017	63.83	84.51	99.16	94.72
2018	72.84	91.83	117.39	99.45
2019	57.16	60.86	108.58	85.89

Source: BP[a].
[a]BP Statistical Review of World Energy 2020.

plants that are typical in the biomass sector. For larger scale power plants, a dedicated supply of biomass fuel is needed. This can be derived from biomass crops, usually fast-growing trees or grasses. Alternatively, some biomass power plant operators rely on biomass pellets, often made from forest wood, which are traded both nationally and internationally.

Biomass wastes are the cheapest of these fuels. Prices will vary but should usually be low, making generation from these fuels economical even in

TABLE 4.6 Price for biomass residuals in the United States in 2020.

Type of residual fuel	Price ($/Tonne)
Wood manufacturing residuals	41.4
Sawmill residuals	36.3
Roundwood/pulpwood	33.7
Other residuals	32.1

Source: Statista[a].
[a]Average price of biomass feedstock in the United States in February 2020, by product, Statista www.statista.com.

relatively inefficient power plants. Table 4.6 shows figures for biomass waste (residuals) in the United States in early 2020 derived from data published by Statista. Waste from wood manufacturing plants was the most expensive at $41.4/tonne, followed by sawmill residuals at $36.3/tonne. Roundwood and pulpwood (wood suitable for papermaking) cost $33.7/tonne and other residuals cost $32.1/tonne. These prices would suggest that waste or residual fuels of this type are relatively competitive with coal.

Annual average monthly prices for biomass pellets in the United States, collated by the US Energy Information Administration, are collected in Table 4.7 for 2016−19. Biomass pellets are manufactured from felled wood and are considered a high-quality, tradable commodity. Large volumes of biomass pellets are exported from the United States and Canada to Europe. The average prices shown in the table are notably higher than for the biomass residuals in Table 4.6. For example, in 2016, the average price was $178/tonne. Average annual prices were not available for succeeding years so monthly average prices for December each year are shown. In 2017, this was $168/tonne; in 2018, it was $180/tonne and in 2019, it was $192/tonne. This

TABLE 4.7 Average US biomass pellet prices.

Year	Average pellet price ($/Tonne)
2016	178
December 2017	168
December 2018	180
December 2019	192

Source: US EIA[a].
[a]Monthly Densified Biomass Fuel Report, 2016−19, US Energy Information Administration.

suggests that biomass pellets are significantly more expensive than coal. Similar prices can be found in Europe where bulk pellet prices were between €170/tonne and €323/tonne in December 2018 according to Bioenergy Europe.[6] Power plants burning biomass can both avoid carbon taxes and benefit from renewable subsidies in some European countries; this can help offset the high price for the fuel.

Locally grown chipped biomass wood fuel is likely to be more competitive than pellets for combustion in a biomass power plant. This type of fuel will typically be supplied on a long-term contract between the company supplying the fuel and the power company burning it. Prices for such contracts are not generally available.

Hydrogen

Hydrogen is beginning to appear as a commercial power plant fuel. It is attractive because, of itself, it is a clean fuel with little environmental impact. The gas can be produced from natural gas, but this process leaves carbon dioxide as a waste byproduct and damages the environmental credentials of the fuel. Industrial production of this type is taking place in Brunei to produce hydrogen from natural gas, the product being exported to Japan for use in a power plant. The Japanese market for hydrogen is expected to expand rapidly during the third decade of the 21st century.

More important from an environmental perspective is the production of hydrogen by the electrolysis of water. If this can be achieved using low-cost renewable electricity, then the resulting fuel gas is truly clean because the only product of its combustion in air is water vapour. The cost of hydrogen generated by electrolysis will depend on the production cost of the electricity used in the process. It has been suggested by the International Renewable Energy Agency (IRENA) that by 2018 it was possible to generate hydrogen for $5−6/kg by connecting a hydrolyser to the Danish electricity grid. Meanwhile, the US Department of Energy has set a target of $5/kg for hydrogen dispensed from pumps, and in Japan, the target is to reduce the pump price of $10/kg in 2018 to $3/kg by 2030.[7]

Nuclear fuel

The cost of a unit of nuclear fuel cannot be directly related to the cost of electricity in the same way as can the cost of a unit of any of the other fuels discussed above. In contrast to the situation for fossil fuel, biomass or

---

6. European Bioenergy Outlook 2019, Pellet, Bioenergy Europe.
7. Hydrogen from Renewable Power: Technology Outlook for the Energy Transition, IRENA, September 2018.

hydrogen, it is not particularly useful to discuss the energy input in terms of a kilogram or a tonne of nuclear fuel.

One kilogram of radioactive U^{235} could, in theory, release 24,000,000 kWh of heat energy while a kilogram of coal produces around 8 kWh. A typical nuclear reactor will contain about 100 tonnes of uranium but only about 2% of this will be U^{235}. A typical 1000 MW nuclear power plant would probably consume around 3 kg of this radioactive uranium each day.

The fuel for a nuclear power plant is contained in technically complex fuel rods, and the nuclear material in each fuel rod is itself the product of a long manufacturing process. It is possible to put a market price on the important constituent of most fuel rods, pure uranium, but that does not provide much guidance when examining the economics of nuclear power.

In spite of these caveats, the nuclear industry does advance figures suggesting that it is possible to compare nuclear power plant fuel costs with the fuel costs for fossil fuel plants. For example, the US Nuclear Energy Institute has estimated that the cost of fuel for a nuclear plant is around 14% of total costs, although this increases to 34% when factors such as waste management and additional front-end costs are included.[8] The fuel cost for a coal-fired plant is, on the same basis, estimated to be 78% of total costs and for a gas-fired power plant it is 87%. Against this, the cost of construction of a nuclear power plant is significantly higher than the cost of construction of a fossil fuel power plant. And the comparison with renewable energy cannot easily be made because in general the fuel cost for a renewable plant is zero.

---

8. Figures are from the World Nuclear Association.

Chapter 5

The capital cost of a power plant

The construction of a power station is a major civil and electrical engineering project, and the total cost of even the smallest generating facility will be an important factor in determining its economic viability. The capital cost will therefore have a major bearing on the choice of power station technology. Capital costs are not static and the relative cost of different types of generating technology will vary over time. This will change the economic balance between them. And while the ultimate determinant of economic viability should be the cost of the electricity the power station produces over its lifetime, it is quite possible that at the time of construction the owner of the plant will choose the least expensive to build.

The capital cost is itself determined by several factors including the sophistication of the generating technology, by the materials involved in its construction and by the work required to build the power station. Some very small power plants, a small piston engine for backup power for example, can be purchased from a factory and delivered ready to plug in and switch on. More usually there will be a significant amount of civil engineering involved too, preparing the site for the power plant and installing the components. Some, such as large hydropower plants, will involve massive civil construction projects. In other cases, such as onshore wind turbines, the site preparation is less complex but installation is a highly skilled process.

The type of power station chosen will have a major effect on the balance of these costs. There are several power plant types that use components that are mass-produced and then delivered to the site. Gas turbines, piston engines and even wind turbines are fabricated in factories and delivered ready to use or ready to assemble. For these factory-produced power plants, the unit cost often depends on global competition between different manufacturers. For other technologies, much of the fabrication takes place at the power station site. Many of the components of a large coal-fired power station will be built at the site, as will a great deal of a nuclear power plant. For these plants, labour costs will be high and the local cost of labour will impinge significantly on the overall capital cost. This will vary from country to country.

The Cost of Electricity. https://doi.org/10.1016/B978-0-12-823855-4.00005-X

Project finance

Then, on top of these other considerations, there will be the cost of financing the project. It is rare today for a power plant to be built without the need for a loan to finance its construction. This loan must eventually be paid down, usually with revenues from the power station. Financing will add another significant component to the overall cost.

Debt financing brings into focus another factor, the length of time over which the loan must be paid down. Most power plants are expected to have operating lives of between 20 and 30 years and the loan repayment will normally be stretched over as much of this lifetime as possible to keep the payments low. It is rare for loan repayments to be longer than 30 years — often they are shorter — but there are types of power plants which have much longer lifetimes. Hydropower plants can often operate for 100 years or more with adequate maintenance. Many nuclear power plants have also had their lives extended to 50 or 60 years, although this may involve additional expense. Well-maintained solar cell power plants may also be able to operate for longer than their nominal lifetimes.

If the operating life of a power plant is much longer than the period over which a financing loan is paid down, this will front-load the cost of the plant making it relatively more expensive than a scheme where the loan repayment matches the plant lifetime. On the other hand, once the debt has been repaid, the cost of electricity from the long-lived power station will drop significantly. There are hydropower plants and nuclear plants operating today, free of debt, that produce some of the cheapest power available.

Project financing, while widespread, can sometimes lead to problems in a world where economic or environmental directions are changing. Of particular concern in the third decade of the 21st century is the prospect that new fossil fuel power stations will become redundant well before the end of their operating life because they are supplanted by renewable energy sources. If a plant is taken out of service before its debt is repaid, it becomes what is known as a 'stranded asset'. Stranded assets become a financial liability for their owners who will try to recoup their losses, perhaps by charging more for electricity from other power stations they own. It is a contentious issue but one that needs to be acknowledged as global warming rises up the international agenda.

Power plant capacity factors

When comparing the cost of different power generating technologies today, it is not enough simply to look at the cost per unit of generating capacity of each. That would be reasonable if each type of generating technology could run for the same amount of time each year, but such is not the case. Fossil fuel and nuclear power plants should be capable of running for most of the time without interruption, if required, but many renewable power plants can only operate intermittently. This difference must be taken into account when comparing one with another.

The parameter that is used to measure the availability of a power plant is called its capacity factor. The capacity factor is simply the number of hours in a year that the plant is able to operate divided by the total number of hours in a year. A power station with a capacity factor of 100% would be capable of running endlessly, while one with a capacity factor of 20% would only be able to operate for 1 h in every five. This latter plant would only be able to produce one-fifth of the electricity of the former in every year. Or, five of these plants would be needed to replace one plant of similar size with a 100% capacity factor.

There is a further level of complexity to consider, the theoretical and the actual capacity factor. The theoretical capacity factor of a power plant is the capacity factor it would achieve if it was able to operate for as long as it was capable. For a conventional fossil fuel or nuclear power plant, this would be limited only by the amount of time it must be taken out of service each year for maintenance (or in the case of a nuclear power plant for refuelling) and the average amount of time lost due to failure of components. A coal-fired power station would be expected to have a theoretical capacity factor of around 85%, and for a natural gas-fired combined cycle power plant, it would be 87%. A nuclear power station might typically achieve 90% capacity factor, similar to that of a geothermal power plant while a biomass-fired combustion plant could be expected to reach a capacity factor of 83%.[1]

Generating plants based on renewable sources of energy will generally have lower theoretical capacity factors. A hydropower plant could in theory have a capacity factor similar to that of a fossil fuel plant, but in practice it is usually significantly much lower and many vary from year to year as a result of variations in annual rainfall. The theoretical capacity factor of wind and solar plants depends both on the technology and on the resource. Solar capacity factors are often below 30%, and for wind it may be 40% or less, but there are exceptions.

Actual capacity factors are another matter. For many fossil fuel power plants today, this will depend on the duty cycle required of the plant. With the advance of renewable energy, gas turbine, some coal and even nuclear plants may be used by grid operators for support rather than base load generation. This will reduce the actual capacity factor compared to that theoretically achievable.

The reverse is often true of renewable resources. Wind and solar power plants are generally dispatched first and so they will often achieve close to the theoretical limit. Of these renewable sources, only hydropower plants are likely to be held in reserve.

---

1. These figures are based on US EIA estimates.

Table 5.1 shows figures for the annual capacity factors of conventional power plants in the United States between 2010 and 2019 based on figures from the US Energy Information Administration (US EIA). These figures illustrate many of the points discussed above. For example, the figures for US nuclear power plants show that they have typical annual capacity factors of over 90% and reached 94% in 2019. This is a reflection of the fact that the US nuclear plants are running flat out providing base load generation. Many of these US plants are over 30 years old and have paid down their construction debt, making the power they produce relatively cheap.

The capacity factors for coal-fired power plants show an (almost) steady decline between 2010 and 2019. While there are times where the capacity factor rose, year upon year, the fall from 67% in 2010 to 48% in 2019 is consistent with a gradual phasing out of coal-fired power in the United States, partly in consequence of environmental considerations but heavily influenced by the availability of cheap natural gas during this period. Supporting this, the annual capacity factor for natural gas−fired combined cycle plants increased from 44% in 2010 to 57% in 2019. While the rate of increase was not monotonic, the trend is consistent with the increased use of cheap natural gas for power generation. This is a uniquely US factor, not mirrored across the globe, although there has been a similar shift from coal to natural gas in many developed nations.

TABLE 5.1 Annual capacity factor for conventional power plants in the United States.

Year	Coal	Natural gas combined cycle	Natural gas open cycle	Gas engine	Nuclear
2010	67.1%	44.3%	7.8%	6.5%	91.1%
2011	62.8%	44.3%	7.9%	8.4%	89.1%
2012	56.2%	52.2%	8.9%	7.3%	86.6%
2013	59.4%	48.8%	8.3%	8.8%	90.8%
2014	60.5%	48.6%	8.3%	10.8%	91.7%
2015	54.3%	55.8%	9.8%	11.9%	92.3%
2016	52.8%	55.4%	11.0%	11.5%	92.3%
2017	53.1%	51.2%	9.6%	11.6%	92.3%
2018	53.6%	55.0%	11.9%	13.0%	92.5%
2019	47.5%	56.8%	11.8%	13.9%	93.5%

Source: US EIA[a].
[a]*Electric Power Monthly, May 2020, US Energy Information Administration.*

The other two columns in Table 5.1 show the annual capacity factor for open cycle natural gas turbines, typically used for peak power production and gas engines, also often used for peak production and grid support. Consistent with this, both show annual capacity factors which are under 12% in most years.

Table 5.2 presents similar sets of figures to Table 5.1 for US renewable energy plants. These figures are for the most part a reflection of the maximum capacity factor of which these technologies are currently capable. For solar photovoltaic (solar PV) power plants, the typical annual capacity factor of around 20% in 2010 had risen to 25% by 2013 and remained in that region for the rest of the decade until 2019. Solar thermal power plants have shown a more variable annual capacity factor ranging from 25% in 2010 to 17% in 2013. This technology depends more critically on the quality of sunlight available and will be more dependent on meteorological conditions.

The output from the US wind farms shows a valuable increase in capacity factor during the decade in the table. From 29% in 2010, the average capacity factor had risen to a consistent 35% by 2019. This is a reflection of the improvement in wind technology. Wood biomass power plants, meanwhile, were consistently operating at around 60% during the whole of the decade. This would suggest that the theoretical capacity factor attributed to these biomass plants of 83% from the US EIA quoted above is rather optimistic.

TABLE 5.2 Annual capacity factor for renewable power plants in the United States.

Year	Hydro	Solar pV	Solar thermal	Wind	Wood
2010	37.5%	20.2%	24.5%	29.7%	61.5%
2011	45.8%	19.0%	23.9%	32.1%	59.6%
2012	39.6%	20.4%	23.6%	31.8%	61.3%
2013	38.8%	24.5%	17.4%	32.4%	59.0%
2014	37.2%	25.6%	18.3%	34.0%	60.0%
2015	35.7%	25.5%	21.7%	32.2%	59.3%
2016	38.2%	25.0%	22.1%	34.5%	58.3%
2017	43.0%	25.6%	21.8%	34.6%	60.2%
2018	41.9%	25.1%	23.6%	34.6%	60.6%
2019	39.1%	24.5%	21.2%	34.8%	60.9%

Source: US EIA[a].
[a]Electric Power Monthly, May 2020, US Energy Information Administration.

The first column in Table 5.2 shows annual capacity factors for US hydropower plants. These range from 36% in 2015 to 46% in 2011. Part of this variability will be a result of variations in annual rainfall in the United States. However, another consideration is the increasing use of hydropower for grid support rather than base load generation. The US geothermal power plants, not shown in the table, had average annual capacity factors of between 68% and 76% in the decade to 2019.

Table 5.3 presents a set of complementary figures for global average capacity factors for renewable technologies assembled by the International Renewable Energy Agency (IRENA). The first column shows global average figures for solar PV. These are notably lower than the US figures, ranging from 14% in 2010 to 18% in 2019. However, the steady improvement over this period once again reflects improvements in solar PV technology. Similarly the average capacity factor for solar thermal plants rose from 30% in 2010 to 45% in 2019. In addition to improvements in the technology, this may be a result of more solar thermal plants around the world being equipped with energy storage.

Globally, hydropower plants showed a larger average capacity factor than the US plants in the previous table. However, the maximum was still only 51%. Bioenergy plants had average capacity factors of between 67% and 86%.

TABLE 5.3 Global average capacity factor for renewable technologies (%).

Year	Solar photo-voltaic	Solar thermal	Onshore wind	Offshore wind	Hydropower	Bioenergy
2010	14.0	30.0	27.1	36.8	43.9	71.7
2011	15.3	35.5	27.7	37.9	44.1	68.1
2012	15.1	27.4	28.5	40.4	45.8	64.5
2013	16.4	31.0	27.0	45.4	50.0	74.3
2014	16.6	28.5	28.8	30.2	49.2	74.7
2015	16.5	40.4	29.1	39.8	50.6	75.5
2016	16.7	36.2	30.6	39.0	50.3	67.2
2017	17.7	38.6	32.3	45.1	47.0	85.8
2018	18.2	45.1	34.0	42.2	45.5	76.4
2019	18.0	45.2	35.6	43.3	48.4	70.0

Source: IRENA[a].
[a]*Renewable power Generation Costs in 2019, IRENA.*

Perhaps the most notable figures in Table 5.3 are for wind energy. The third column shows average annual capacity factors for onshore wind. This rose steadily during the decade shown on the table, from 27% in 2010 to 37% in 2019. Again, this is a reflection of the improvement in the wind technology being deployed over the period of the table. Column four shows similar figures for offshore wind. These are more variable, with a minimum of 30% in 2014 and a maximum of 45% in 2013 and 2017. These figures are consistently higher than similar figures for onshore wind and show to extent to which the offshore resource is superior to that onshore.

Figures published by Statistica for UK offshore wind farms, one of the largest national fleets, showed an average annual capacity factor in 2019 of 40.6%,[2] in line with the figure in Table 5.3. Notable, however, was the performance of one Scottish offshore wind farm — as reported by Energy Numbers — which achieved a record average annual availability of 53.3%.[3]

Capital costs

The cost of building a new power station will vary from project to project. Even seemingly identical projects may have different costs and the prospective cost will only emerge when detailed planning is carried out. There are a range of important situations, however, when some idea of cost is needed before detailed planning begins. To meet this need, a variety of national and international organisations regularly calculate the expected capital cost of power plant construction for a range of different types of technology. Important series of this type are published by the International Energy Agency (IEA), by the US EIA, by Lazard, by the IRENA, by trade bodies for specific technologies such as solar power or wind power and by government agencies in many parts of the world. The figures from these studies can be used by project planners to help select a specific technology suitable to a particular situation. An examination of the year-on-year changes in these cost figures can also reveal important trends that can influence policy and decision-making.

In order to be able to compare the cost of a coal-fired power plant with that of a wind farm, or the cost of a hydropower plant with that cost of a nuclear power plant, the costs for each must be reduced to a standard unit of measure. In this case, the most common approach is to present the cost for each kilowatt of generating capacity. This figure will allow direct comparisons to be made between technologies, and by scaling up to the size of a proposed project, a rough estimate of actual total cost can be obtained. The figures are normally calculated

---

2. *Offshore wind energy load factors in England in 2017 and 2018, by region,* Statistica.
3. *UK offshore wind capacity factors,* 31/01/2020, Energy Numbers.

before any cost for financing a project and are taken into consideration. This figure is sometimes called the overnight cost. All the figures in the tables below are overnight costs for plant construction.

Table 5.4 presents figures produced by Lazard for the capital cost of all the main generating technologies in the United States in 2019. These figures provide an annual snapshot of the expected cost in the United States but in many cases will be broadly transferrable to other regions and nations.

The figures in the table offer cost ranges and the most expensive of these is nuclear power which was estimated to cost between $6900/kW and $12,200/kW. Of all the established technologies such as coal, natural gas and nuclear power plants that have historically been used for base load generation, new nuclear power plants have the highest price, by a significant margin.

There are three fossil fuel—based generating technologies included in the table. The most expensive of these is coal-fired technology which had an estimated capital cost in the range $3000/kW to $6250/kW in 2019. A natural gas-fired combined cycle plant, the most common technology for large-scale utility gas-fired power generation, costs between $700/kW and $/1300/kW or roughly four to five times less than a similarly sized coal plant. The third fossil fuel technology, an open cycle gas turbine, had an estimated cost of $700/kW to $900/kW. These latter units are usually used for providing power during periods of peak demand.

TABLE 5.4 Capital cost of US generating technologies, 2019.

Technology	Cost ($/kW)
Coal	3000–6250
Natural gas combined cycle	700–1300
Open cycle gas turbine	700–950
Nuclear	6900–12,200
Utility-scale solar photovoltaic (PV)	900–1100
Rooftop solar PV	1750–2950
Solar thermal with energy storage	6000–9100
Onshore wind	1100–1500
Offshore wind	2925
Geothermal	3950–6600

Source: Lazard[a].
[a]*Lazard's Levelied Cost of Energy Analysis — Version 13.0, Lazard.*

Six renewable technologies and configurations are also included in Table 5.4. The most expensive of these is solar thermal power generation with energy storage with a cost range of \$6000/kW to \$9100/kW. Geothermal generating capacity had an expected cost in 2019 of \$3950/kW to \$6600/kW. Onshore wind power, which is now one of the most important renewable generating technologies, globally, costs between \$1100/kW and \$1500/kW in the United States, while offshore wind capacity could be built for \$2925/kW. The cheapest renewable technology in terms of capital cost was utility-scale solar generation, which had a price of \$900/kW to \$1100/kW. Rooftop solar deployment is more expensive with a price range of \$1750/kW to \$2950/kW, but this is deployed at the consumer or distribution system level where the cost of electricity is also higher.

Capital cost trends for all the most important technologies will be discussed in the sections below, but Table 5.5 shows figures from IRENA for the average global annual total installed cost for the main renewable technologies. These figures are based on actual power plants. The table contains figures for both hydropower and bioenergy, neither of which is included in Table 5.4.

TABLE 5.5 Global average total installed cost for renewable technologies (\$/kW).

Year	Solar photo-voltaic	Solar thermal	Onshore wind	Offshore wind	Hydropower	Bioenergy
2010	4702	8987	1949	4650	1254	2588
2011	3936	10,588	1939	5326	1236	1302
2012	2985	8183	1972	4741	1321	1595
2013	2615	6419	1828	5738	1494	3028
2014	2364	5510	1781	5245	1361	2982
2015	1801	7361	1642	5260	1491	2592
2016	1637	7737	1635	4280	1784	2175
2017	1415	7324	1628	4683	1824	2897
2018	1208	5253	1549	4245	1435	1693
2019	995	5774	1473	3800	1704	2141

Source: IRENA[a].
[a]Renewable power Generation Costs in 2019, IRENA.

The global 2019 capital costs in Table 5.5 show broad agreement with the US figures in Table 5.4 for most of the technologies that appear in both. The only exception is offshore wind, which has an average global capital cost in 2019 of $3800/kW, based on IRENA's database of projects, compared with $2925/kW for the US EIA estimate. Some of the trends in Table 5.5 appear erratic, but for the new renewable technologies, wind and solar, the cost in 2019 is lower than it was in 2010, and in the case of solar PV, the price drop is dramatic.

The two established renewable technologies, hydropower and bioenergy, show no obvious trends in Table 5.5. Prices for each rise fall sharply again and then rise once more. For hydropower, the average installed cost rose from $1254/kW in 2010 to $1704/kW in 2019, a price rise that is probably consistent with inflationary increases over that period. The bioenergy capital costs are much more erratic, ranging from $1300/kW in 2011 to $2028/kW in 2013. This variability is probably a result of the small number of such projects built each year. More consistent figures from the United States are presented below and these put the cost of a new plant in 2019 at around $4000/kW.

Coal plant capital costs

Coal-fired power plants are normally large, complex facilities. New generating stations will generally employ the most efficient technology available based on what are known as supercritical or ultra-supercritical boilers. These boilers burn pulverised coal to generate high-temperature, high-pressure steam. The temperatures and pressures involved necessitate the use of advanced materials which are more expensive than conventional steels. Moreover, a coal-fired power station will also require a range of emission control facilities to clean the exhaust gases from the boiler before they can be released into the atmosphere. Electricity from such a plant is generated using a large steam turbine generator. Typical capacities for new plants range from 700 to 1500 MW and efficiencies from 41% to 51%.[4] The largest plants can have generating capacities in excess of 4000 MW.

Many of the components of a coal-fired station will be fabricated and erected at the site. This will incur high labour costs so the cost of labour in the region where the plant is built will be an important consideration. Further, the use of large quantities of materials such as concrete and steel will make costs vulnerable to swings in global commodity prices. Construction times are long, typically 3−4 years. No major coal-fired power stations have yet been built with carbon capture and storage (CCS) but that may become a requirement in future years. If so, it will increase the capital cost further as well as reducing plant efficiency.

---

4. *Projected Costs of Generating Electricity, 2015 Edition*, IEA, NEA and OECD.

Table 5.6 shows the estimated overnight capital cost of coal-fired power generation in the United States between 2000 and 2019 based on figures from the US EIA. The estimated cost of advanced coal technology was $1021 in 2000 and then rose slowly during the first 5 years of the decade, reaching $1167/kW in 2005. From that point, prices rose sharply over the next 5 years so that in 2010 they were estimated to be $2625/kW, an increase of 125% compared to 5 years earlier. Much of this rise can be attributed to a massive

TABLE 5.6 Annual overnight capital cost for US coal-fired power plants ($/kW).

Year	Pulverised coal	Pulverised coal with CCS
2000	1021	n/a
2001	1046	n/a
2002	1079	n/a
2003	1091	n/a
2004	1134	n/a
2005	1167	n/a
2006	1206	n/a
2007	1434	n/a
2008	1923	n/a
2009	2076	n/a
2010	2625	n/a
2011	2658	n/a
2012	2694	4662
2013	2734	n/a
2014	2726	n/a
2015	n/a	4649[b]
2016	n/a	5072
2017	n/a	5132
2018	n/a	5212
2019	3661	5851

Source: EIA[a].
[a]Assumptions to the Annual Energy Outlook, 2000–2020, US Energy Information Administration.
[b]A pulverised coal plant with 30% carbon capture and storage (CCS).

increase in global commodity prices in the middle and late part of the first decade of the century, a trend that was brought to an abrupt halt by the global financial crisis of 2008. In consequence, the estimated cost of new coal-fired capacity stabilised again, rising to $2726/kW in 2014 — a rise of only 4% in 4 years.

The US EIA capital cost series for unabated coal plants stopped between 2014 and 2018, only resuming in 2019 when the estimated cost for a new plant was put at $3361/kW — a 23% increase over the price in 2014. In 2012, the US EIA also began to produce estimates for a coal plant with CCS.[5] That year, the cost was put at $4662/kW, a 73% premium over the cost of a similar plant without CCS. By 2019, the cost estimate for a coal plant with CCS was $5851/kW, or 60% higher than the unabated plant.

The IEA, Nuclear Energy Agency (NEA) and Organisation for Economic Co-operation and Development (OECD) have together produced a series of reports charting the projected cost of generating electricity across the globe. The most recent of these five yearly reports, *Projected Costs of Generating Electricity, 2015 Edition* (IEA report) was published in 2015. The capital costs recorded in this report are all for advanced coal-fired stations. For OECD countries, the cost for a new plant ranged from $1218/kW in South Korea to $3067/kW in Portugal. Of the two non-OECD nations reporting costs, the overnight cost in China was put at $813/kW, while in South Africa, it was $2222/kW. Most of the OECD costs quoted are in line with those in the United States between 2010 and 2015 in Table 5.6. However, both South Korea and especially China are outliers. It is possible that cost in both nations is a reflection of government subsidies supporting local industries and thereby depressing commodity and labour prices.

Natural gas–fired power plant costs

Natural gas–fired power stations come in a variety of configurations, but the most common types likely to be built in the third decade of the 21st century are based around gas turbines. The most efficient and the most effective of these stations are combined cycle gas turbine plants which can be used for base load generation or, increasingly, for grid support services. Simpler open cycle gas turbines are also widely used, in this case usually for peak power generation at times of high demand.

All gas turbines are technically advanced prime movers and they are manufactured by a limited number of companies globally. There is often intense competition between these companies and that has led to exceptionally low prices for some turbines. Combined cycle gas turbines are manufactured in

---

5. Previous to this the US EIA produced estimates for the cost of an integrated coal gasification combined cycle power plant with carbon capture and storage. This is not considered a competitive technology today.

factories and then shipped to the power plant site, essentially complete. Installation, while still complex, is therefore much simpler than for a coal plant and on-site preparation will be simpler too. The time from order to completion is around 3 years. Simple cycle gas turbines have much smaller generating capacities than combined cycle units and easier to install. Lead times for plants based on these are typically 2 years.

Natural gas—fired power plants are cleaner than coal-fired plants and the emission control systems they require are simpler. No large gas turbine plant has been built with CCS, but this will likely become a requirement during the third decade of the century.

Table 5.7 contains figures for the estimated overnight capital cost of three types of gas turbine power plant in the United States between 2000 and 2019. The combined cycle power plant in the table is an advanced, high-efficiency flexible unit. For this type of station, the cost in 2000 was estimated to be $533/kW. The cost both rose and fell between 2000 and 2005 when it was $532/kW, virtually identical to the price 5 years earlier. The global boom in the last years of the first decade then pushed up the cost to $917/kW by 2010, a rise of 72% in 5 years. Prices subsequently stabilised again so that although there were small rises in the middle of the next decade, the estimated cost in 2019 was $954/kW, only 4% higher than in 2010.

Open cycle gas turbines are generally somewhat cheaper than the combined cycle variants and this can be seen in Table 5.7. Here the estimated cost of an open cycle turbine in 2000 was $440/kW, but by 2005, this had fallen to $367/kW. Prices jumped somewhat in the last part of that decade, reaching $626/kW in 2010, and then stabilised again. In 2019, the estimated cost for an advanced open cycle gas turbine power plant was $710/kW, 13% higher than a decade earlier.

In 2003, the US EIA started to estimate the annual capital cost for an advanced combined cycle power plant with CCS. That year it put the cost at $969/kW, 70% higher than for the plant without CCS. The estimated cost of this configuration rose towards the end of the decade and then stabilised during the succeeding decade. In 2019, the estimated cost was $2470/kW, nearly 160% more costly than the plant without CCS.

Figures for the capital cost of combined cycle power plants from the 2015 IEA report put the price for plants built within OECD nations at between $845/kW in South Korea and $1289/kW in New Zealand with many plants being built at a cost similar to those in Table 5.7 for the period between 2010 and 2015.[6] The efficiencies of the cited plants ranged from 45% to 60%. However, the cost in China, the only non-OECD nation included, was $627/kW. As with coal-fired plants, South Korea and China are again outliers. For open cycle gas

---

6. *Projected Costs of Generating Electricity, 2015 Edition*, IEA and NEA.

TABLE 5.7 Annual overnight capital cost for US gas-fired power plants ($/kW).

Year	Advanced combined cycle	Advanced combined cycle with carbon capture and storage	Advanced open cycle
2000	533	n/a	440
2001	546	n/a	451
2002	563	n/a	439
2003	569	969	444
2004	517	992	356
2005	532	1021	367
2006	550	1055	379
2007	654	1254	450
2008	877	1683	604
2009	897	1720	617
2010	917	1813	626
2011	929	1834	634
2012	931	1833	632
2013	945	1856	641
2014	942	1845	639
2015	1000	1898	632
2016	1013	1917	640
2017	1026	1936	648
2018	736	1963	658
2019	954	2470	710

Source: EIA[a].
[a]*Assumptions to the Annual Energy Outlook, 2000–2020, US Energy Information Administration.*

turbine power stations, the cost variation within the OECD was from $500/kW in the United Kingdom to $933/kW in Belgium.

Nuclear power plant costs

Nuclear power stations are probably the most technically complex power plants used to generate electricity commercially in the 21st century. These plants have a radioactive core where the nuclear reactions that generate heat occur. This core is surrounded by a containment vessel that will retain any radioactive material in case of a plant failure. The heat generated in the core is used to make steam to drive a large steam generator. The use of radioactive materials requires that nuclear power plants have extensive safety features to prevent failure or in the case of failure to shut the plant down without harm. These safety features are expensive to implement.

Construction of a nuclear power station will involve massive civil engineering work at the site and large volumes of concrete and steels will be consumed. Some of the components will be manufactured off-site, but much of the fabrication will take place at the site of the station. These plants have the longest lead times of any power project, typically 6 years; some advanced stations have taken much longer to complete.

Nuclear power plants are usually large. Minimum unit capacity is rarely below 1000 MW and a unit size of 1600 MW is not uncommon. When two or more units are constructed at a single site, capacity can easily exceed 3000 MW. This size, combined with the technical complexity required to maintain safety, makes the construction of a nuclear power station the most capital intensive of any type of power project.

Table 5.8 contains figures for the estimated annual overnight capital cost of a new advanced nuclear power station in the United States. In 2000, the estimated cost was $1729/kW. Between 2000 and 2005, estimated costs moved both up and down so that in 2005 the cost was put at $1744/kW, a less than 1% increase in 5 years. Prices rose steeply between 2005 and 2010, again reflecting increasing commodity prices, so that by 2010 the estimated cost was $4567/kW, an increase of 162%. Costs stabilised during the succeeding 5 years before starting to increase again in the middle of the second decade, and by 2019, the estimated cost was $6016/kW, a rise of 32% in 9 years. This makes nuclear power the most expensive of all the established generating technologies.

Global costs are generally similar to those observed in the United States, with some notable exceptions. The IEA report found the cost of nuclear plants in OECD nations built or started between 2010 and 2015 to be between $2021/kW in South Korea and $6216/kW in Hungary. The cost of a nuclear plant in China was between $1807/kW and $2615/kW. Again the two Asian nations are outliers. Cost variations will often depend on the differences in the regulatory regimes covering nuclear power in the different nations or regions.

Fuel cell power plants

Fuel cells are technically advanced devices, similar in concept to a battery, that have the ability to generate electricity from gaseous hydrogen. There are several different types of fuel cell. Most are available in modular form, with

TABLE 5.8 Annual overnight capital cost for US nuclear power plant ($/kW).

Year	Advanced nuclear
2000	1729
2001	1772
2002	1750
2003	1669
2004	1694
2005	1744
2006	1802
2007	2143
2008	2873
2009	3308
2010	4567
2011	4619
2012	4700
2013	4763
2014	4646
2015	5288
2016	5091
2017	5148
2018	5224
2019	6016

Source: EIA[a].
[a]Assumptions to the Annual Energy Outlook, 2000–2020, US Energy Information Administration.

sizes ranging from a few kW for domestic use to hundreds of kilowatts for commercial use. Large power plants are constructed by installing multiple units.

All fuel cells are fabricated in factories and then shipped to the site where they are to be used. This makes installation relatively simple, particularly for smaller units. They have low environmental impact and this allows them to be used in urban settings if necessary. Some of the smaller units are particularly suited to combined heat and power applications. Efficiency is high when a fuel cell uses hydrogen as fuel, but this fuel has limited availability so many use natural gas instead. This is converted into hydrogen before use. Many fuel cell types require catalysts such as palladium metal to operate and this makes them expensive. While economies of scale should help reduce prices over the long term, production volumes are not sufficiently high for this to have much effect yet. In consequence, fuel cells are still relatively expensive.

Table 5.9 shows the estimated US EIA annual overnight capital cost for fuel cells in the United States between 2000 and 2019. At the start of this

TABLE 5.9 Annual overnight capital cost for US fuel cell power plant ($/kW).

Year	Fuel cell
2000	1767
2001	1810
2002	1850
2003	1872
2004	3679
2005	3787
2006	3913
2007	4653
2008	4640
2009	4744
2010	5846
2011	5918
2012	6045
2013	6099
2014	6042
2015	6217
2016	6252
2017	6192
2018	6250
2019	6671

Source: EIA[a].
[a]Assumptions to the Annual Energy Outlook, 2000–2020, US Energy Information Administration.

series, the cost of a fuel cell power plant was $1767/kW. However, by 2005, this had risen to $3787/kW, and in 2010, it reached $5846/kW, 231% higher than 10 years earlier. Much of this steep rise in costs can be put down to commodity prices, particularly for the rare metals needed as catalysts. Prices stabilised somewhat after 2010, and by 2019, the prices was $6671/kW, an increase of only 14% in 9 years.

The capital cost of a fuel cell is high compared to many competing technologies. However, the devices are often deployed at the distribution system level where wholesale electricity prices are higher.

The capital cost of hydropower

Hydropower is the oldest and best established renewable technology. The technology is usually subdivided into two types, small hydropower and large hydropower. Large hydropower plants, often with dams and reservoirs, are massive civil engineering projects with concomitant price tags. These schemes can provide water for irrigation and for drinking as well as electric power and may often be subsidised by governments or local authorities. Small hydropower schemes are less disruptive and cheaper to build, but their unit capital cost is often higher than for a large project. Costs are very site-specific too, so schemes of identical generating capacity may have markedly different capital costs.

A major hydropower scheme will often have a generating capacity of several hundred megawatts, with the largest reaching over 1000 MW. Smaller projects can range in size from 1 or 2 megawatts to 20 −30 MW. Much smaller schemes are possible too, and generators with capacities of 1 kW or less are common.

The major electromechanical component for a hydropower plant is the turbine generator. These devices are fabricated in factories and then shipped to the project site. However, most of the construction work involved in a hydropower scheme must take place at the site. In consequence, the capital cost will depend critically on local labour costs. Where a large dam is constructed, there may also be sensitivity to the price of concrete.

Table 5.10 shows figures from the United States for the estimated overnight capital cost of a medium- or large-scale hydropower scheme from 2006 to 2019. The US EIA did not produce estimates for this technology between 2000 and 2005. The capital cost of a hydropower plant was put at $1364/kW in 2006. By 2010, this had risen to $2019/kW. There was a gradual increase in the estimated cost during the succeeding 9 years, and by 2019, the capital cost was estimated to be $2752/kW.

The 2015 IEA report contains overnight costs for hydropower plants across the globe. For large hydropower schemes, the costs reported varied between

TABLE 5.10 Annual overnight capital cost for US hydropower plant ($/kW).

Year	Hydropower
2006	1364
2007	1410
2008	2038
2009	2084
2010	2019
2011	2134
2012	2179
2013	2213
2014	2410
2015	2191
2016	2220
2017	2634
2018	2680
2019	2752

Source: EIA[a].
[a]Assumptions to the Annual Energy Outlook, 2000–2020, US Energy Information Administration.

$2933/kW in Portugal for a 144 MW plant and $1567/kW for an 800 MW project in Brazil. Smaller projects were relatively more expensive. A 12 MW scheme in Japan costs $8687/kW, while a 10 MW project in Switzerland had a capital cost of $6848/kW. In China, meanwhile, the cost of a 13,050 MW hydropower scheme was $598/kW.

Biomass power plant costs

Biomass power plants are combustion plants that use similar technology to that in a coal-fired power station. Most biomass plants are much smaller than coal-fired plants and the technology is often rather more primitive. A biomass plant will typically have a capacity of 30 MW or less, although larger plants have been constructed. This makes them relatively less efficient that most coal plants do. The combustion process will generate some nitrogen oxides and carbon dioxide and the former may need control. It is unlikely that carbon capture will be added to a conventional small biomass combustion plant but it has been mooted for one or two large coal-fired power stations that have been converted to biomass.

The first column of Table 5.11 shows US EIA figures for the annual overnight capital cost of a biomass combustion plant in the United States between 2000 and 2019. The capital cost in 2000 was estimated to be $1464/ kW, and like capital costs for other technologies in this chapter, the price rose

TABLE 5.11 Annual overnight capital cost for US biomass power plants ($/kW).

Year	Biomass	Landfill gas
2000	1464	1308
2001	1536	1336
2002	1569	1365
2003	1588	1381
2004	1612	1402
2005	1659	1443
2006	1714	1491
2007	2490	1773
2008	3339	2377
2009	3414	2430
2010	3395	7698
2011	3519	7694
2012	3685	7858
2013	3590	7751
2014	3399	7730
2015	3498	7954
2016	3540	8059
2017	3584	8170
2018	3642	8313
2019	4080	1557

Source: EIA[a].
[a]Assumptions to the Annual Energy Outlook, 2000–2020, US Energy Information Administration.

slowly during the first half of the succeeding decade before accelerating so that by 2010 it had reached $3395/kW, a rise of 132% in 10 years. Costs stabilised during the next decade, before eventually rising slowly so that in 2019 the estimated cost was $4080/kW.

Figures from the IEA report for biomass combustion plants between 2010 and 2015 indicate that a 100 MW plant in the United States costs $4587/kW. Meanwhile, the addition of co-firing of wood pellets to a coal-fired power plant in the Netherlands cost $587/kW, while in the United Kingdom, it cost $719/kW.

In addition to biomass combustion technology, Table 5.11 also contains figures for the cost of biomass electricity production from landfill gas. This is based on gas engines, reciprocating engines that burn the biogas generated from landfill sites. This represents a small part of the biomass energy sector, but one that is easy to exploit.

The US EIA figures for landfill gas-based generation in the second column of Table 5.11 contain unexplained anomalies. The cost of the technology in 2000 was estimated to be $1308/kW and this rose slowly until 2006 and then more rapidly so that in 2009 the estimated cost was $2377/kW. However, in 2010, this rose, without explanation, to $7698/kW, or three times the cost 1 year earlier. Prices remained in this region until 2018 when the cost was estimated to be $8313/kW, but in 2019, the price fell, again without explanation, to $1557/kW.

Other overnight costs from the IEA report can add some perspective to these US EIA figures. In Italy, the cost of a plant burning biogas in a gas engine was $8667/kW, close to the EIA figures between 2010 and 2015. However, the cost for the same technology in Spain was between $1852/kW and $3773/kW. The limited sample of projects makes it difficult to draw and clear conclusions for this technology.

Geothermal power plant costs

Geothermal power plants use heat extracted from underground reservoirs of hot brine to drive a heat engine and generate electric power. Normally, the brine is pumped to the surface and then used to produce low-pressure steam that will drive a steam turbine. The low pressure and relatively low temperature of the steam produced results in a geothermal energy conversion process that has low efficiency. Even so this can be economically viable because the energy source is free.

Table 5.12 shows estimates for the overnight capital cost of geothermal power technology in the United States between 1999 and 2019. For the first decade of the century, these figures are rather erratic, starting at $1626/kW in 1999 and rising to $2960/kW before falling back to $1057/kW in 2007.

TABLE 5.12 Annual overnight capital cost for US geothermal power plants ($/kW).

Year	Geothermal
2000	1626
2001	1663
2002	1681
2003	2099
2004	2960
2005	2100
2006	1790
2007	1057
2008	1630
2009	1666
2010	2364
2011	2393
2012	2444
2013	2375
2014	2331
2015	2559
2016	2586
2017	2615
2018	2654
2019	2680

Source: EIA[a].
[a]Assumptions to the Annual Energy Outlook, 2000–2020, US Energy Information Administration.

However, from 2010 to 2019, the series shows more stable prices, from $2364/ kW in 2010 to $2680/kW in 2019.

The IEA report contains overnight capital costs for geothermal technology in other parts of the world between 2010 and 2015 which show a wide spread. For example in Turkey a 24 MW plant cost $1493/kW, while in New Zealand the cost for a 250 MW power plant was $3331/kW. In the United States, during this period, an 80 MW plant cost $6291/kW and a 90 MW plant cost $5992/kW. The capital cost of geothermal technology is extremely site-specific and this will account for the wide variation in the cost of actual plants reported by the IEA.

The capital cost of wind power

Wind power is one of the important new renewable technologies. Total global installed capacity has grown rapidly during the 21st century and the technology has become more robust and reliable. Modern wind turbines are manufactured by a limited number of key companies around the world which sell their products internationally. This has created global competition which has helped force prices down. However, there are also regional factors which affect prices.

There are two principal branches of the wind turbine market, onshore wind turbines and offshore wind turbines. The use of onshore wind turbines is widespread, but the main offshore market is in Europe, and particularly in the United Kingdom. The size of turbines for offshore use continues to increase, helping reduce the unit capital cost, but installing very large turbines onshore is often limited by the ability to transport the units to the site. Most wind turbines are installed as part of a wind farm comprising an array of turbines operating as a single power station. All modern wind turbines are manufactured in a factory and then assembled on site. The main components are usually made from steel or special composite materials, so manufacturing costs will depend at least partly on commodity prices. However, standardisation and the use of multiple units in wind farms helps keep the cost of installation to a minimum.

Table 5.13 shows US EIA figures for the annual overnight capital cost of wind power in the United States. The estimated cost for onshore wind turbines in 2000 (column one of the table) was $919/kW, but by 2010, this had risen to $2251/kW, an increase of 145% in 10 years. This mirrors the rise in cost for other technologies during the first decade of the century. Prices stabilised around 2010 and then the effect of global competition began to take hold so that during the second decade of the century prices began to drop. By 2019, the estimated cost for onshore wind power in the United States was $1319/kW, a fall of just over 40% in 9 years. This fall in capital cost has made onshore wind power one of the most competitive sources of electric power in the United States.

The second column of Table 5.13 contains a similar series of costs for offshore wind power in the United States between 2007 and 2019. Offshore wind is more expensive than onshore wind because of the much higher cost of installation. In 2007, the cost of offshore wind was estimated to be $2547/kW

TABLE 5.13 Annual overnight capital cost for US wind power plants ($/kW).

Year	Onshore wind	Offshore wind
2000	919	n/a
2001	918	n/a
2002	938	n/a
2003	949	n/a
2004	1060	n/a
2005	1091	n/a
2006	1127	n/a
2007	1340	2547
2008	1797	3416
2009	1837	3492
2010	2251	4404
2011	2278	4345
2012	2032	4452
2013	2061	4505
2014	1850	4476
2015	1536	4605
2016	1576	4648
2017	1548	3952
2018	1518	4758
2019	1319	4356

Source: EIA[a].
[a]Assumptions to the Annual Energy Outlook, 2000–2020, US Energy Information Administration.

or 90% higher than for an onshore wind installation. Prices rose to a peak of $4648/kW in 2016, after which the estimates become somewhat erratic. The estimated price in 2019 was $4356/kW.

Figures from the IEA report for the early part of the second decade of the 21st century suggest that the price of onshore wind in nations of the OECD up to 2015 was between $1571/kW in the United States and $2999/kW in Japan. The equivalent capital cost in China was between $1200/kW and $1400/kW. Offshore wind capital costs varied between $3703/kW in the United Kingdom and $5413 in France. However, given the trend shown in Table 5.13, it is likely that the cost for both onshore and offshore wind would be significantly lower in 2019.

Solar power capital costs

Solar power is the second important new renewable generating technology alongside wind power, and like wind power, the installed capacity based on solar power has soared during the 21st century. The principal type of solar power generation is based on solar (PV) cells, solid state devices that convert sunlight directly into electricity. The solar converters are manufactured in high volumes using solid state fabrication techniques, and this has allowed enormous economies of scale to be achieved. While the manufacture of these devices is limited to a number of high-technology companies, the market for solar cells has become global and this has also resulted in extremely competitive pricing, particularly during the second decade of the century.

Table 5.14 contains, in the second column, UA EIA figures for the estimated annual overnight cost for utility-scale solar PV power plants in the United States between 2000 and 2019. The cost in 2000 was $3681/kW. This climbed to $5879/kW in 2009 before economies of scale and the effect on global markets of cheaper solar cells manufactured in China started to depress prices everywhere. From 2010 to 2019, the cost of solar cells has fallen year upon year, and in 2019, the estimated capital cost in the United States was $1331/kW. This dramatic fall — the cost in 2010 was over three times higher than in 2019 — has had an equally dramatic effect on the market for solar cells.

The table does not show the cost of rooftop solar cells which are generally more expensive than the large arrays for utility-scale generation. However, Table 5.4 shows both and indicates a cost that is two to three times higher than the utility array cost. Meanwhile, global overnight costs published in the IEA report indicate that costs in other parts of the world were much lower than in the United States. In fact, the figures for solar cells in Table 5.14 look overly pessimistic.

There is a second type of solar power plant called a solar thermal plant. These use the sun as a heat source to drive a heat engine and so share similarities with more conventional power plants. Solar thermal plants rely on arrays of mirrors which are costly to manufacture and to install. The energy conversion technology can be novel and expensive too, though some use conventional steam cycles. The global capacity of this type of generation is relatively small and costs remain high compared with many other technologies.

TABLE 5.14 Annual overnight capital cost for US solar power plants ($/kW).

Year	Solar thermal	Solar photovoltaic
2000	2394	3681
2001	2157	3317
2002	2204	3389
2003	2478	3810
2004	2515	3868
2005	2589	3981
2006	2675	4114
2007	3499	5380
2008	4693	5750
2009	4798	5879
2010	4333	4474
2011	4384	4528
2012	4653	3624
2013	4715	3394
2014	3787	3123
2015	3895	2362
2016	3908	2169
2017	3952	2004
2018	4011	1876
2019	7191	1331

Source: EIA[a].
[a]Assumptions to the Annual Energy Outlook, 2000–2020, US Energy Information Administration.

The first column of Table 5.14 shows estimated annual capital cost for solar thermal technology. The cost in 2000 was estimated to be $2394/kW, but this had risen to $4333/kW by 2010. Costs stabilised somewhat during the succeeding decade, peaking at $4715/kW in 2013 before falling back slightly, but in 2019, the US EIA revised its estimate, pushing the cost up sharply, to $7191/kW.

These figures may be compared with figures from the IEA report for overnight costs from the early part of the second decade of the century, between 2010 and 2015. For two US solar thermal plants, the installation cost was $3561/kW and $4901/kW, while a plant constructed in Spain during this period had an installed cost of $8142/kW. Solar thermal technology should benefit from economies of scale, but it is unlikely that it will ever compete directly on cost with solar PV technology.

Chapter 6

Lifecycle analyses for fuels and power generation technologies

The capital cost of a power station, discussed in the previous chapter, provides a metric against which to compare different types of power generation technology. Of itself it provides a limited indication of the economic value of a particular power plant, but it is one of the key inputs into another metric called the levelized cost of electricity (LCOE). The LCOE is the output from an economic model that attempts to assess the present, and future, cost of electricity from a power station that has not yet been constructed. The model is an example of what is known as a lifecycle analysis, in this case of the financial inputs and outputs from a power station from its construction and through its life until final decommissioning.

The LCOE model adds together the cost of building a power plant, the cost of financing any loans required to facilitate its construction, the cost of operating the power plant over its full lifetime and the cost of any fuel it uses during that period, the cost of maintenance and repair and finally the cost of decommissioning and dismantling the power plant once its usefulness is exhausted. This total lifetime cost of the power station is then compared with the total amount of electricity that the power plant will actually produce over the same lifetime to arrive at a cost for each unit of power the plant generates.

This type of calculation could be carried out once the power plant has been decommissioned but that would be of limited use. What is required is a cost today for a power plant that has not yet been built. To achieve this, the model makes a variety of assumptions about future conditions in order to arrive at a meaningful future cost of electricity. These assumptions may lead to inaccuracy and bias, but the model does provide a metric for comparing the cost of electricity from different technologies. The LCOE for different technologies will be explored more fully in the final chapter of the book, but the concept is examined here because it is an important example of lifecycle analysis as applied to power plants.

The LCOE model is an economic model dealing specifically with financial costs. There are a range of other lifecycle analyses that can be applied to power station technologies in order to illuminate other aspects of their performance. For example, by using a similar approach, lifecycle analysis can show how much carbon dioxide each type of power station will emit over its lifetime for

The Cost of Electricity. https://doi.org/10.1016/B978-0-12-823855-4.00006-1

each unit of electricity it generates while similar analyses can be applied to other pollutant emissions such as sulphur dioxide, nitrogen oxides or particulates. A lifecycle emission analysis will add together the amount of pollutant produced during the manufacture of the materials needed to build the power plant, emissions during its construction, further emissions that take place while it is operating and emissions that are consequent on its decommissioning a dismantling. These emissions are then added together and divided by the total number of units of electricity the plant produces over its lifetime to provide an emission rate per unit of electricity.

Another group of lifecycle analyses considers the energy performance of power plants. Energy conversion efficiency is a simple metric of this type; another type of energy analysis examines the time it takes for a power plant to generate the amount of energy used in its construction. This is closely related to the energy payback ratio for a power plant, how much energy a plant gives back compared with the energy invested in it. Similar to this is the energy return on investment (EROI), which can take a broader view. EROI has recently been used as part of a wider analysis of the value of energy to society and how its value influences the ability of societies to support various activities.

Many of these metrics change over time. For example, the EROI will change for fossil fuel power plants as it becomes more difficult, and more expensive in energy terms, to recover the fossil fuel from the earth. Or, the greenhouse gas emissions per unit of power from a solar cell will fall when the energy used to make the actual solar cell material is produced from renewable sources rather than fossil fuels.

Boundary conditions

Lifecycle analyses can provide valuable insights into the relative performance of different technologies, particularly in areas other than economic, but their results depend critically on how the limits or boundary conditions of each analysis is defined. Boundary conditions are necessary because a power station (or any other object of lifecycle analysis) is part of an interacting system we call the world. So, if we want to calculate how much energy a power station uses over its lifetime, we will include any energy consumed transporting fuel to the plant. But should we consider it appropriate to include the energy each power station worker consumes when preparing breakfast each morning?

There are clear arguments against extending the boundaries of a study this far. Breakfast is a part of everyday life whether one is working or not. In other cases, the choice of boundary might be more arbitrary. For example, when calculating the energy inputs for an analysis of nuclear power, the boundary conditions may include the cost of refining uranium from uranium ore but omit the energy cost of mining and transporting the ore, and it almost certainly will not include societal cost associated with the labour required to mine and transport the ore. Or in the case of wind power the boundary conditions might include the energy cost of building the foundations for the plant, but the analysis may choose to exclude the energy required to manufacture the concrete that was used in the foundation construction.

One of the most contentious boundary discussions centres around externalities such as the cost to society of pollutant emissions from power plants. Putting a cost on these factors is difficult, and fossil fuel power plants have, traditionally, not taken them into account when lifecycle economic analyses are carried out. The LCOE lifecycle analysis discussed above rarely includes the environmental cost of different generating technologies. However carbon emission costs may be included. Taxing carbon emissions is a way of attempting to include some of these external factors associated with global warning but even this is unlikely to account for the full cost. This is discussed more fully in Chapter 8.

As will be clear by now, by careful choice of boundary conditions it is possible to skew the results of a lifecycle analysis to present one technology in a better light than others. But even when the drawing of the boundaries is attempted objectively, there can still be a systematic bias. It is important, therefore, to ensure that boundary conditions are clearly and transparently defined.

Lifecycle energy analyses

One of the most important areas in which lifecycle analysis can be applied to power generation is energy balance — how much energy comes out of a power station compared with the amount that goes in. Perhaps the simplest of this type of calculation is the energy conversion efficiency of the power station.

Energy conversion efficiency is an important thermal power plant parameter because it provides a measure of the efficiency of the energy conversion process, how much of the input energy — the fuel — is actually converted into useful electricity. So, for example, an older coal-fired power station with a subcritical boiler might achieve an energy conversion efficiency of 38% while a modern supercritical plant can push this to 45%. More advanced ultra-supercritical boilers can raise this to 48% and may in future achieve 50% or higher. Clearly the higher the efficiency, the more electricity is produced from each unit of coal, and this will improve both the economic and the environmental performance of the plant.

The boundaries of the analysis that produces this type of figure are very tightly drawn, and they are normally restricted to the actual power station itself. The main aim is to show how much electricity is generated for each unit of energy that enters the plant. To achieve this, the amount of fuel or energy entering the plant is compared with the amount of electrical energy that leaves.

Table 6.1 shows the energy conversion efficiencies for all the most common generating technologies in use today. Looking at thermal technologies

TABLE 6.1 Energy conversion efficiency of power generating technologies.

Technology	Typical best case energy conversion efficiency
Natural gas combined cycle	60%
Pulverised coal power plant	45%
Diesel engine	50%
Solar thermal	16%
Solar photovoltaic	20%
Wind power	40%
Nuclear power	33%

Source: Author's own figures.

first, the most efficient is the natural gas—fired combined cycle power plant with an efficiency of 60%. Diesel engines are relatively efficient too at 50%, while modern coal-fired power plants typically achieve 45%. Nuclear power is a thermal technology too. Large nuclear stations operate at around 33% energy conversion efficiency.

For renewable technologies, the comparison is generally less favourable. A modern wind farm might be able to reach 40% efficiency in converting wind energy into electricity while a solar photovoltaic plant will probably only reach 20%. Solar thermal technology is even less efficient; 15% efficiency is a typical figure for a plant of this type.

There is a snag here, however. When a wind turbine captures energy from the wind and converts it into electricity, the energy it does not capture continues on its way, as wind. However, when a coal-fired power plant burns a tonne of coal, converting 45% of the energy it contains into electricity, the 55% of the energy not converted into electricity has still been consumed and will emerge from the plant as waste heat. This heat is released into the environment and may be considered an additional emission. So to compare the energy conversion efficiency of a combined cycle plant to that of a wind turbine is not to compare like with like. And while the energy conversion efficiency figures for the different technologies are useful from an electro-mechanical point of view, comparing them in this way across technologies is not particularly illuminating in most contexts.

Energy payback ratio

A more useful metric can be obtained if the amount of energy contained within the fuel or energy source is removed from the analysis. The cost in energy terms of mining and transporting a fuel and the cost of cleaning up any emissions are included, but the quantity of energy that enters the power station, be that contained in sunlight, wind or natural gas, is not included. What is of interest here is the amount of energy emerging from the power plant as electricity compared to the amount of energy spent building and operating the plant (including any fuel mining and transportation energy costs).

In this case, the 'birth to death' analysis of the energy performance of a power plant tries to show how much energy is actually provided to society by the operation of the power station, distilling the performance into a single parameter. This may be expressed as an energy payback time, the amount of time it takes for the power plant to generate energy equivalent to that required to built and operate it or by factoring in the lifetime of the power plant it can be expressed as an energy payback ratio showing the amount of energy the power plant delivers over the course of its lifetime for each unit of energy spent.

Table 6.2 shows a set of energy payback ratios published by the Canadian utility Hydro Quebec in 2004. Bearing in mind that Hydro Quebec operates a large fleet of hydropower stations, it may not be surprising to find that hydropower is by some margin the best performing technology in the table, with an energy payback ratio of 170—180. These figures may be somewhat optimistic, but hydropower plants do have extremely long lifetimes if well

TABLE 6.2 Energy payback ratios for power generation technologies.

Power geneneration technology	Energy payback ratio
Hydropower	170—280
Wind power	34—18[1]
Biomass generation	3—27
Nuclear power	14—16
Solar photovoltaic	3—6
Pulverised coal	2.5—5.1
Natural gas combined cycle	2.5—5
Pulverised with carbon capture and storage	1.6—3.3

Source: Hydro Quebec[2].

1. The Hydro Quebec study concluded that offshore wind had a payback ratio as low as −18. A more recent study, greenhouse gas emissions and energy performance of offshore wind power, Hanne Lerche Raadala, Bj°rn Ivar Volda, Anders Myhrb and Tor Anders Nygaard, Renewable Energy, June 2014, pp. 314—324, put the energy payback ratio of offshore wind at between 7.5 and 12.9.
2. Electricity Generation Options: Energy Payback Ratio, Hydro Quebec, 2004.

designed and most analyses place them among the best performing technologies on this metric. The next highest performing technology in the table is wind power with a payback ratio of up to 34 for onshore wind but as low as −18 for offshore wind. More recent analysis shows offshore wind in a much more favourable light. However, the payback ratio for solar photovoltaic generation is only 3−6. This is a reflection of the high energy cost for the manufacture of the single crystal silicon needed for high-efficiency solar cells. Nuclear power has an energy payback ratio of 14−16 based on this analysis.

Combustion technologies fare less well. Biomass combustion has an energy payback ratio of 3−27, but this higher figure is probably for the combustion of waste biomass fuel. A pulverised coal-fired power plant and a natural gas−fired power plant both show similar ratios (2.5−5.1 and 2.5−5.0, respectively) while a pulverised coal plant with the addition of carbon capture and storage pushes this down to between 1.6 and 3.3. Bearing in mind that a ratio of less than one indicates that a power plant consumes more energy over its lifetime than it actually generates, the latter figures are not encouraging for this technology.

These figures present renewable technologies in a favourable light compared with fossil fuel technologies. Other analyses, discussed below, can present then in a different light.

Energy return on investment

EROI has, in the last decade, become a more favoured way of presenting this type of lifecycle energy analysis. While the result is essentially the same as the energy payback ratio discussed above, recent studies have looked in much more detail at the boundary conditions used to calculate the EROI, at the way in which values vary from country to country and at the way in which EROIs vary over time. In addition to this, a broader analysis of EROI and energy costs has been applied to societies in general to provide insight into how the cost of energy affects the economic wealth and well-being of a society or nation.

One way of looking at the economic implications of energy is to calculate the annual monetary cost of energy to a nation − the amount it costs to buy all the energy needed in a year − and then compare this with the annual national gross domestic product.[3] According to published observations, when the value of this ratio is around 5%, which is considered low, then an economy can grow strongly and it can afford to invest in such areas as scientific research and can support artistic endeavours. However, a rise to between 10% and 14% is

---

3. This analysis is based on EROI of different fuels and the implications for society, Charles A.S. Hall, Jessica G. Lambert and Stephen B. Balogh, Energy Policy, 2014, Volume 64, pp. 141−152.

usually linked to an economic recession. Such rises are often found during periods of energy price shock such as sudden rises in the cost of oil.

Table 6.3 presents figures for the EROI of a variety of common fossil fuels and power generation technologies from a paper published in Energy Policy.[4] These figures are based on a meta-analysis of figures published in a large number of other studies. The figures for the fuels in the table do not necessarily relate to electricity production. They simply compare the energy spent making the energy source available to society with the energy that the source actually provides — the heat energy released when natural gas is burned, for example.

The mean EROI for the production of oil and gas in Table 6.3 is 20:1. A broader examination of the EROI for these fuels over time suggests that the value is gradually falling.[5] For example, the value in 1995 was estimated to be around 30:1, but by 2006, it had fallen to 18:1. This appears to be related to the increased cost of both prospecting for oil and gas and recovering it. Alternative sources of oil and gas such as tar sands and oil shale tend to show much lower EROI values, with a typical value of 4:1 as shown in Table 6.3.

The EROI for the other major fossil fuel, coal, is relatively high at 46:1. Its value appears to remain relatively high over time, although there are changes

TABLE 6.3 Energy return on investment (EROI) for power generation fuels and technologies.

Fuel/Technology	Mean EROI
Oil and gas	20:1
Coal	46:1
Tar sands and oil shale	4:1
Ethanol	5:1
Nuclear power	14:1
Hydropower	84:1
Wind power	18–20:1
Solar photovoltaic	10:1
Geothermal	9:1

Source: Energy Policy.[6]

4. EROI of different fuels and the implications for society, Charles A.S. Hall, Jessica G. Lambert and Stephen B. Balogh, Energy Policy, 2014, Volume 64, pp. 141−152.
5. EROI of different fuels and the implications for society, Charles A.S. Hall, Jessica G. Lambert and Stephen B. Balogh, Energy Policy, 2014, Volume 64, pp. 141−152.
6. EROI of different fuels and the implications for society, Charles A.S. Hall, Jessica G. Lambert and Stephen B. Balogh, Energy Policy, 2014, Volume 64, pp. 141−152.

observable. For example, one study found that the EROI for coal in the United States was around 80:1 in the 1950s but fell to 30:1 in the mid-1980s and then rose again to around 80:1 by the 1980s.[7] The variation in the EROI is likely to be related to the ease of recovery of coal, including extended use of surface mining which is less costly.

A figure for the EROI for ethanol is also included in Table 6.3. The mean figure quoted is 5:1, but many of the studies from which this mean was calculated were lower than this, with one or two high outliers pushing the mean up, the authors note. This may suggest the true EROI for the biomass fuel is actually lower than 5:1.

The EROI figures for combustion fuels in Table 6.3 cannot all be compared directly with those in Table 6.2 for combustion technologies because the latter refers exclusively to use of fuels for electricity generation while Table 6.3 figures include other uses such as for transportation fuel. However, for other types of generation, the figures are comparable because the product in each case is electricity.

Most nuclear power plants produce electricity alone. For nuclear power, the mean EROI from the studies examined was 14:1. This matches very closely to the figure quoted in Table 6.2. However, other studies (see below) arrive at much higher figures for nuclear technology.

The technology with the highest EROI, by some margin in Table 6.3, is hydropower. The figure quoted in the table is 84:1, which, while significantly lower than the figure in Table 6.2, is close to twice the value for the nearest rival. However, the EROI for hydropower varies from site to site and values might be much lower.

Of the other renewables, wind power has an EROI of 18:1 to 20:1. For solar photovoltaic, the EROI quoted in Table 6.3 is 10:1. This latter ratio is much higher than in Table 6.2. However the higher figure is reliant on a weighting for the high-quality power generated by solar cells. Unweighted estimates are often closer to the 3:1 ratio from the earlier table. Finally, geothermal energy has an EROI of 9:1.

Table 6.4 provides figures from another study, published in Energy,[8] that looked exclusively at the energy return for power generation. In this case, the figure for natural gas generation is 28.0, higher than the estimate for oil and gas as a fuel, while that for coal is 20.0, much lower than in the previous table. However, brown coal is given a much higher figure, 31.0, probably consistent with the fact that brown coal is usually surface mined and the power plant is often adjacent to the mine, both of which lower the energy associated with

7. Aggregation and the role of energy in the economy, Cutler J Cleveland, Robert K Kaufman and David I Stern, Ecological Economics Vol 32, February 2000, pp. 301–317.

8. Energy intensities, EROIs (energy returned on invested), and energy payback times of electricity generating power plants, D Wei_bach, G Ruprecht, A Huke, K Czerski, S Gottlieb and A Hussein, Energy, 1 April 2013, Volume 52, pp 210–221.

TABLE 6.4 Energy return on investment (EROI) and energy payback times for electricity generating plants.

Electricity generating technology	EROI	Energy payback time
Natural gas combined cycle	28.0	9 days
Biomass-fired combined cycle	3.5	12 days
Solar photovoltaic (polycrystalline)	3.8–4.0	6 years
Solar thermal (parabolic trough)	21.0	1 year
Wind energy	16.0	1 year
Hydropower	50.0	2 years
Coal-fired power plant	20.0	2 months
Brown coal–fired power plant	31.0	2 months
Nuclear power	75.0	2 months

Source: Energy.[9]

recovery and transportation. Other figures in the table are broadly consistent with those in the earlier table with the exception of nuclear power which is assigned an EROI of 75.0 here. The exceptional figure appears to be a result of a variety of changes to the assumptions for nuclear power including increases in the plant lifetime and in the number of hours of operation each year. Such variations from study to study again emphasise that it is important to examine the assumptions and boundary conditions when looking at lifecycle analyses of this sort.

The study from which the figures in Table 6.4 were taken also attempted to calculate a value for the EROI of the main renewable technologies when account is taken of their variability and the need, therefore, for some form of backup to support them. This served to reduce the overall EROI of all these technologies. For example, the value for solar thermal technology from the table, 21.0, was reduced to 9.6 using this assumption. For wind energy, the reduced EROI was only 4.0 while for hydropower it was reduced from 50.0 to 35.0 and for solar photovoltaic generation it was reduced to 2.3.[10]

Table 6.4 also contains estimates for the energy payback time for different generating technologies. These vary from 9 days for a natural gas–fired

9. Energy intensities, EROIs (energy returned on invested) and energy payback times of electricity generating power plants, D Wei_bach, G Ruprecht, A Huke, K Czerski, S Gottlieb and A Hussein, Energy, 1 April 2013, Volume 52, pp. 210–221.

10. Energy intensities, EROIs (energy returned on invested), and energy payback times of electricity generating power plants, D Wei_bach, G Ruprecht, A Huke, K Czerski, S Gottlieb and A Hussein, Energy, 1 April 2013, Volume 52, pp 210–221.

combined cycle plant to 6 years for a solar cell. As with all the figures quoted in Tables 6.2, 6.3 and 6.4, the values depend critically on the assumptions and boundary conditions. Change one or both of these and the results will change. So while there is some consistency between the three tables, there is also a great deal of variability.

Lifecycle greenhouse gas emissions

Another important type of lifecycle analysis shows how much of a pollutant gas is emitted by a power plant over its lifetime for each unit of electricity the plant produces. The boundary conditions for this type of study tend to be more clearly defined than for EROI studies and the results are generally more consistent. However, the values will change, particularly for combustion technologies when efficiencies improve.

This type of study is important because it will provide a direct comparison of the environmental impact of power generation technologies on the environment. There are several important pollutants that can be studied in the way. They include carbon dioxide, sulphur dioxide, nitrogen oxides and particulates.

TABLE 6.5 Lifecycle greenhouse gas emissions for electricity generation.

Technology	Lifetime greenhouse gas emissions (gCO_2 equivalent/kWh)
Coal	980
Natural gas combined cycle	450
Natural gas open cycle	670
Biomass (short rotation wood crop)	45
Geothermal	11–47
Nuclear	17
Wind	11
Solar photovoltaic (crystalline)	45
Solar thermal	23
Ocean power	16
Hydropower	4

Source: NREL.[11]

11. Figures are taken from various sources from the US National Renewable Energy Laboratory LCA Harmonization web page: https://www.nrel.gov/analysis/life-cycle-assessment.html.

Table 6.5 presents figures for the lifecycle greenhouse gas emissions from the main generation technologies. The figures in the table are presented as grams of carbon dioxide equivalent for each kilowatt hour of electricity generated (gCO_2 equivalent/kWh). This unit is used because there are a range of gases that can cause greenhouse warming of the atmosphere in addition to carbon dioxide. The most important of these is methane, which is released in smaller quantities, but is more potent than carbon dioxide. Methane dissipates from the atmosphere more quickly than carbon dioxide too, but over a 20 year time frame it is 84 times more potent. This reduces to 28 times over 100 years. Other potent greenhouse gases include nitrous oxide, which is around 260 times more potent over both time frames.

As might be expected, the largest greenhouse gas emissions are from coal-fired power plants. Based on studies by the US National Renewable Energy Laboratory (NREL) shown in the table, coal plants emit 980 gCO_2 equivalent/ kWh. Coal is primarily composed of carbon so its main combustion product is carbon dioxide.

An open cycle gas turbine burning natural gas typically emits 670 gCO_2 equivalent/kWh while a natural gas–fired combined cycle plant emits 450 gCO_2 equivalent/kWh. The lower figure for the combined cycle plant is due to its much higher energy conversion efficiency. Natural gas plants produce less carbon dioxide than coal plants, but the production of natural gas leads to significant releases of methane into the atmosphere, which affect the overall environmental performance.

Biomass power plants produce typically 45 gCO_2 equivalent/kWh when burning a wood fuel that is grown specially for the purpose. A plant of this type will produce large quantities of carbon dioxide during combustion, but when the wood crop is regrown, it absorbs some of that carbon dioxide again, leading to the low emission performance.

Nuclear power, which is often considered a low emission technology, emits 17 gCO_2 equivalent/kWh based on NREL analysis. Geothermal power, also based on thermal technology, emits between 11 gCO_2 equivalent/kWh and 47 gCO_2 equivalent/kWh depending on the type of underground reservoir. Some of the latter may release carbon dioxide or methane during geothermal operation.

Of the principal renewable technologies, the best performing is hydro-power with 4 gCO_2 equivalent/kWh. Some hydro schemes can produce methane from the organic material that is submerged when a reservoir is created, but if this happens, it will normally subside over time. Wind power typically produces 11 gCO_2 equivalent/kWh and ocean power around 16 gCO_2 equivalent/kWh. Solar thermal power plants generate slightly more, 23 gCO_2 equivalent/kWh, but the emission for solar photovoltaic plants is higher at 45 gCO_2 equivalent/kWh. This is a result of the electrical energy needed to manufacture the pure silicon which often comes from fossil fuel power stations.

The other pollutants that are released into the atmosphere during electricity production are also the result of fossil fuel combustion. The emission of sulphur dioxide is linked to coal combustion and depends on the amount of sulphur in the coal. Today, this is normally removed from flue gases before they are released into the atmosphere. However, emissions can be as high as 1360 kg/GWh, even with flue gas desulphurisation.[12] Plants that burn heavy fuel oil may also release large amounts of sulphur dioxide.

Coal, natural gas-fired and diesel power plants will all produce nitrogen oxides. There are various technologies that can be used to control these emissions and more advanced plants tend to produce less of each. For example, a coal plant in the United Kingdom without technology to remove nitrogen oxides released 2200 kg/GWh, while a plant with removal technology released 700 kg/GWh according to the World Energy Council figures.[13] All these plants also produce carbon-based particulates. These are usually removed from coal plants together with dust in the flue gases, but emissions can be up to 9800 kg/GWh. Plants burning other fuels generally produce much lower levels of particulates, but emissions from small diesel engine plants without control can be high.

The equivalent lifecycle emissions of these pollutants from nuclear and renewable technologies generally depend on how various materials used in their construction were made. With the exception of biomass combustion, these technologies do not produce any emissions during electricity generation.

12. Comparison of Energy Systems Using Life cycle Assessment, World Energy Council, 2004.
13. Comparison of Energy Systems Using Life cycle Assessment, World Energy Council, 2004.

Chapter 7

Structural issues

One of the biggest issues facing the electricity sector today is to be able to increase the use of renewable energy while maintaining electricity system stability. All the main renewable technologies, hydropower, wind power and solar power, are to differing degrees, intermittent and unpredictable. Both of these characteristics lead to uncertainty regarding the amount of power available. However a power system must always have sufficient reliable power available to meet demand.

There are a number of ways that power system stability can be managed in this situation. The simplest, but potentially the most expensive is to maintain sufficient fast-acting standby capacity that can cut in when demand exceeds supply. The only way this can be achieved today at the scale required is with fossil fuel power plants, generally based on gas turbines burning natural gas. This is the traditional method of maintaining supply and demand in balance, as was discussed in Chapter 3.

Another way of overcoming the problem is to install more renewable generating capacity than is required to meet demand and then invest heavily in energy storage. By this means, surplus power from renewable generators can be stored so that it will be available to use if the renewable generation falls below demand. Most energy storage technologies are relatively fast-acting. The weakness of this approach becomes apparent when renewable generation fails over a long period of time and the stored energy is all used. In addition, energy storage systems are relatively expensive.

Another useful tool for managing supply and demand is demand management. If some grid demand can be shut down when supply levels are marginal, then the need for additional generation can be avoided. However demand management must be capable of being controlled at the grid system operator level with some electricity users agreeing to reduce their demand on request. There are a range of smart grid technologies that can facilitate this type of functionality including the use of smart meters that are in direct communication with the system operator and can receive instructions to control various loads as demand levels vary.

System stability is not only a matter of the quantity of power available to meet demand but also of the quality of that power. There are certain types of grid event that can cause fluctuations in the grid frequency and grid voltage. In traditional grid systems, the size and speed of these fluctuations will be

The Cost of Electricity. https://doi.org/10.1016/B978-0-12-823855-4.00007-3

flattened by the inertia of the generating units connected to the system. Large steam turbines and hydro turbines carry massive rotational inertia and this helps smooth fluctuations. However, much of the new renewable generation is based on either wind or solar power. Wind turbines have relatively small inertia, and solar cells are essentially isolated from the grid because they generate DC power which is converted electronically to alternating current, and these electronic systems do not have any physical momentum or inertia.

All of these means of managing system stability have financial implications. The cost of supplying them is often invisible, but it is nevertheless important. Some power generating technologies rely on these services to a larger extent than others, and this has an impact on the cost of the electricity they produce. This chapter outlines some of the technologies available and the services they can provide.

Peak power

The level of demand on a power system fluctuates continually and this fluctuation must be balanced on the supply side. Small changes in demand can be met relatively easily by allowing the output of base load stations to fluctuate, or by modulating the output from renewable plants. There are also much larger daily fluctuations that cannot be managed in this way, such as the daytime peak in demand typical on most power systems.

These larger fluctuations must be met by having additional capacity available that can be brought on line quickly as demand rises. The traditional approach to this has been to provide the grid with quick responding open cycle gas turbine units that can ramp up and down rapidly as required. These units are relatively inefficient, and they are costly to operate but provide the required additional capacity to maintain grid stability. More recently, larger combined cycle gas turbine plants have also been adapted so that they can respond to demand changes. These plants are less agile than open cycle gas turbines but can provide a good degree of flexibility. Coal-fired and nuclear power plants are also being adapted for flexible operation but they are much less agile than combined cycle plants.

Another approach to managing peak power is to use energy storage plants. When nuclear power was introduced in the 1950 and 1960s, these plants were designed to operate continuously at full output, night and day. However, as nuclear units became larger, this base load operation could lead to surplus power on the grid when demand was low, particularly at night. To manage this, many systems with large nuclear plants also invested in pumped storage hydropower plants that could absorb and store the excess output from these large generation units. This power was then available to help manage the daytime peak in demand. In addition, these hydropower storage plants were extremely fast-acting, so they could increase grid stability. However, storage plants of this type are costly to build and today they are often difficult to finance.

There are many other types of energy storage technology, and in the last two decades, several of these technologies have been introduced at different grid levels to help manage demand. For example, flywheel storage systems are being used to provide extremely fast-acting backup for companies with mission-critical computer systems in case of failure of the grid supply. But, as with pumped storage hydropower, cost is an issue. Tariffs that pay operators of plants that can offer such grid support can help make energy storage economically viable.

Intermittency

The converse of demand fluctuations is fluctuations in the supply side of the system. In traditional grid systems, this was usually the result of the failure of a generating unit — an outage — or a fault in the grid. However, in modern systems that absorb large volumes of renewable power, a variation in supply is a constant feature that is made more significant by the need to dispatch renewable energy first when it is available.

Managing fluctuations in the supply side of the grid relies on the same types of technology that are used to manage peaks in demand, but the unpredictability of these supply side fluctuations makes it much more difficult to manage them. Energy storage systems offer one of the best solutions since they can cut in quickly when needed. Another valuable asset for managing renewable unpredictability is traditional hydropower capacity based on dam and reservoir plants. Like storage plants, these power stations can be brought on line and throttled back rapidly and they offer a cheap way of managing fluctuating output from other plants. However, this capacity is only available when there is water in the reservoir of the hydropower plant, so the capacity must be managed carefully.

The alternative is to use conventional fossil fuel plants, typically open cycle and combined cycle gas turbines. Modern examples of the latter are often designed to be maintained in a parked state so that they can be started rapidly and many have been modified to provide rapid ramping of output, both up and down. One drawback of this mode of operation is a fall in efficiency and an increase is emissions of all types. And for the future, zero emissions can only be achieved by capturing carbon dioxide emitted from such plants and sequestering it so that it cannot reach the atmosphere.

Smart grid technologies

Smart grid technologies bring the functionality of computers and communication systems to bear on the power supply network. There are a number of ways in which smart grid technologies can be used to help maintain system stability. One of the most important is through automated demand management. By setting up rapid communication systems that link the grid operator to the consumers, signals can be passed from one to the other to control the loads that are connected at different times.

At an industrial level, certain large consumers will agree to a reduced tariff with the condition that if demand begins to outstrip supply they will temporarily shut down some or all of their operations in order to reduce the system load. This type of demand management can be fully automated with shutdown activated at different trigger points. However, in many cases, the companies involved will require notice of an impending cutoff in order to shut down in a controlled manner.

Similar control can, in principle, be introduced at the domestic level too, allowing a considerable level of control over demand. One way that this can be implemented is through smart domestic meters which communicate directly with the system controller. These meters, in turn, have control of certain types of domestic appliance such as washing machines or air conditioning systems. With two-way communication, the system control centre can ask these appliances to shut down temporarily when demand is high. Equally importantly though, some devices may also be asked to switch on when there is a surplus of supply over demand, as for example when there is excess wind power on a system.

Smart technologies can also help with the supply side of the grid. One simple application is the use of advanced weather forecasting to predict the output of wind and solar plants attached to the grid. With these forecasts in place, the system controller can schedule additional capacity to come online when output from these renewable plants is expected to be low and plan to take these additional units off line when output rises again. By providing a longer time frame for this type of scheduling to take place, forecasting can reduce the cost attached to adding or removing capacity.

Another useful tool is the virtual power plant. With sufficient communication capacity, it is possible, for example, to aggregate wind plants from different geographical locations and operate them as if they were a single power plant. While wind is unpredictable, the level of unpredictability is much smaller when averaged over a large geographic area. A virtual wind farm of this type will therefore provide a much more reliable output than a single wind farm at a specific location. Moreover, different types of power plant can be aggregated: wind and solar plants are often complementary over a long time scale, for example, with wind output greater in the winter while solar output is higher in summer. The more reliable the output from these virtual plants, the more valuable the power and the higher the price that can be charged for delivery of the power.

Spinning reserve and system stability

The frequency and voltage of a modern grid must be controlled within narrow bands in order for the system to be stable and for consumers to be able to operate their own loads reliably. In a traditional grid system, a major component of the stability was contributed by the large turbine generators connected to the system with their large rotational inertias. These massive devices help smoothen short-term fluctuations.

With the growth of renewable capacity based on wind and solar power, a significant part of this grid inertia has been lost because these renewable units do not present the grid with the same level of rotational inertia. It is possible to compensate for some of this loss with fast-acting energy storage systems, particularly superconducting magnetic energy storage. However, this is insufficient to replace all that has been lost.

The alternative is to pay power plants with large turbine generators to stay on line with their turbines spinning so that they can continue to provide system inertia as needed. This 'spinning reserve' is usually maintained in order to provide the grid operator with fast access to additional power but it can also be used simply to provide grid stability.

The largest source of this type of spinning reserve is from nuclear and fossil fuel plants, particularly coal-fired stations and combined cycle plants. It can also be provided by large hydropower turbines, either in conventional hydropower plants or in pumped storage hydropower stations. Spinning reserve is one of a range of ancillary services that power plant operators can provide to the grid and they offer new ways of gaining revenue from plants that might otherwise be made redundant by the advance of renewable power.

Hydrogen

Hydrogen is a fuel that could potentially provide a solution to several of the problems of grid stability outlined above. The reason for this is that hydrogen can be produced directly from electricity. In particular, in the context of a world that is struggling with global warming, it can be made using surplus renewable electricity from wind and solar power plants. With sufficient renewable capacity to operate in this way, any excess power above that demanded by the grid can be used to make hydrogen which is then stored. When demand rises, the power directed to hydrogen production is reduced, helping keep the grid in balance.

Once it has been made and stored, the hydrogen can be used to generate more power. It can be burned in a gas turbine or in a conventional boiler and it can be used as fuel in a fuel cell. Because the hydrogen produced in this way is 'clean' it can be produced and then burnt without adding to the greenhouse gas load in the atmosphere. Produced in large enough quantities, it can potentially be transported in pipelines to other locations and it can also be used as vehicle fuel.

There are a number of barriers to achieve this type of hydrogen economy, but the technologies needed to achieve it are beginning to be put in place.

Chapter 8

Distorting factors: subsidies, externalities and taxes

There are a number of factors that can distort the economics of electricity production. These factors come in various forms. The last chapter discussed the issue of grid stability and the need for support for intermittent renewable technologies. This support comes at a cost to the system. The additional cost must be added to the nominal cost of production from the renewable generators to provide a true cost of production from these sources. Otherwise the cost of renewable electricity will appear artificially low in comparison to that from other sources.

Another issue that many observers would consider the source of major distortion is that of externalities. In economics, an externality is an effect that an action by one party has on a second party, the cost of which to the second party is not priced into the cost of the activity to the first. Externalities can be both positive and negative. Within the power sector, the most significant set of externalities relate to the harm caused by the combustion of fossil fuels through damage to the environment, through damage to human health and through the effects of global warming. For the most part, the cost of this damage is not priced into the cost of electricity generation from coal, gas or oil. In other words, the polluter causes the damage but the cost of reparation falls elsewhere.

One way of attempting to price in the environmental effect of different types of power generation is by imposing a penalty on those activities that can cause such damage. Measures include the cap-and-trade systems that have been introduced to control carbon dioxide emissions in the European Union (EU) and in other parts of the world, and in the application elsewhere of carbon taxes which tax each unit of carbon dioxide released.

There is a third, very straightforward, means of distorting the cost of energy, through subsidies which directly affect the price consumers pay for their energy. Subsidies take a variety of forms, from preferential grants or discounts for certain types of fuel to direct government subsidies that reduce the cost of electricity to particular consumers. The largest part of global energy subsidies is directed towards fossil fuels today, but renewable generation technologies also benefit from support in many countries too. Subsidies lead to an artificial price for energy that is usually below the actual cost of production.

The Cost of Electricity. https://doi.org/10.1016/B978-0-12-823855-4.00008-5

Subsidies

Subsidies are widespread within the energy industry, and subsidising the cost of fuel or energy is often used as a political tool. In some developing nations with large fossil fuel reserves, the cost of gasoline and electricity to consumers may be subsidised in order to encourage and maintain support for the regime in power. In others, there may be social tariff subsidies that are targetted to help the poorer sections of the population. In developed countries, this type of tariff support is less usual, but subsidies to support particular industries such as coal, nuclear power or renewable energy are not uncommon, and these may, again, have a political incentive. Globally, the larger part of these subsidies relates to fossil fuels.

In the past two decades, as the threat of global warming has become more and more acute, the issue of subsidies for fossil fuels has been highlighted by many of the world's international agencies such as the International Energy Agency (IEA), the World Bank (WB) and the International Monetary Fund (IMF).

Subsidies which reduce the cost of fuel to a consumer lead to increased consumption of the fuel, and in the case of fossil fuels, this leads to larger carbon dioxide emissions as well as more emissions of a range of other harmful pollutants. International agencies such as the IEA, the WB and the IMF have therefore campaigned to encourage countries to reduce these subsidies. This can be politically difficult. If a nation's consumers have become used to low energy prices, then raising the cost will often lead to political unrest. The result is that, often, subsidies are reintroduced. However, the historically low energy prices in 2019 followed by the global pandemic of 2020 and the resulting depressed energy costs may offer a unique opportunity to start to eliminate them.

Table 8.1 presents figures for cumulative global fossil fuel subsidies in 2019, as collated by the IEA. These figures are estimated by using what is known as the price-gap methodology, which compares the average price paid by consumers in each nation to a reference price for the full cost of supply. The difference between the two is then the subsidy level.

According to the IEA analysis, the largest fossil fuel subsidies were directed towards oil with global subsidies of US$150bn in 2019. This was followed by electricity where the subsidy level was US$115bn, natural gas with subsidies of US$50bn and coal with US$2.5bn. Taken together, these figures show total fossil fuel subsidies in 2019 of US$317.5bn. According to the IEA, this represented a fall in overall annual subsidies from 2018 of US$120bn, putting the total annual subsidy in 2018 at US$437.5bn. The fall noted by the IEA from 2018 to 2019 is mostly accounted for by the large drop in oil subsidies as a result of the lower cost of oil products during 2019.

TABLE 8.1 Global fossil fuel subsidies by energy source, 2019.

Energy source	Subsidy (US$ billion)
Coal	2.5
Natural gas	50.0
Electricity	115.0
Oil	150.0
Total	317.5

Source: IEA[a].
[a]*Energy subsidies: tracking the impact of fossil fuel subsidies https://www.iea.org/topics/energy-subsidies.*

Meanwhile, the International Renewable Energy Agency (IRENA), which takes the IEA figures as a starting point but applies a broader approach, calculated that total fossil fuel subsidies in 2017 were US$447bn. Again the subsidies for oil-based products were the highest at US$220bn, followed by electricity at US$128bn.

Table 8.2 presents a breakdown of subsidies by nation for the 25 countries with the largest level of subsidies, based again on IEA analysis. By far the highest level of subsidies was found in Iran which underwrote national fossil fuel purchases with US$86bn in 2019. Of these subsidies, the largest part, US$51.7bn, was accounted for by electricity subsidies with the remainder relatively evenly split between gas and oil. The nation with the next highest subsidy level was China with US$30.5bn in subsidies, US$18.1bn for oil and US$12.4bn for electricity.

Virtually all the countries with significant levels of fossil fuel subsidy are fossil fuel—producing nations. For example, Saudi Arabia had the third highest level of subsidies in 2019 at US$28.7bn, mostly supporting the use of oil, while the fourth nation, Russia, provided subsidies of US$24.1bn equally divided between electricity and natural gas. India (US$21.9bn), Indonesia (US$19.2bn), Egypt (US$16.4bn), Venezuela (US$12.7bn) and Iraq (US$7.4bn) made up the rest of the top 10 by subsidy level. Of all the nations in the table, only one, Kazakhstan, provides a significant level of subsidy for coal. However some of the countries listed, particularly China, will be providing a subsidy for coal combustion through their subsidising of the cost of electricity.

The IEA has also estimated the total level of subsidies for each nation as a proportion of its gross domestic product (GDP). On this basis the outliers are Iran where subsidies are 18.8% of GDP, Venezuela with 16.7% of GDP

TABLE 8.2 Fossil fuel subsidies in 25 top nations, 2019.

Country	Oil (US$ billion)	Electricity (US$ billion)	Gas (US$ billion)	Coal (US$ billion)
Iran	18.0	51.7	16.3	—
China	18.1	12.4	—	—
Saudi Arabia	18.2	5.8	4.7	—
Russia	—	13.7	10.4	—
India	21.0	—	0.9	—
Indonesia	19.2	—	—	—
Egypt	9.1	6.4	0.4	—
Algeria	8.8	2.0	2.3	—
Venezuela	7.1	4.5	1.1	—
Iraq	5.9	1.3	0.2	—
Kazakhstan	3.1	1.0	0.3	2.2
UAE	—	0.6	5.0	—
Kuwait	1.2	3.1	1.2	—
Libya	3.7	0.7	—	—
Uzbekistan	0.3	1.4	2.7	—
Argentina	3.2	0.2	1.0	—
Mexico	—	3.3	—	—
Turkmenistan	1.0	0.3	1.8	—
Ecuador	3.0	—	—	—
Ukraine	—	2.2	—	—
Azerbaijan	1.0	0.5	0.4	—
Pakistan	—	—	1.7	—
Malaysia	1.8	—	—	—
Nigeria	1.7	—	—	—
Bangladesh	—	0.9	0.8	—

Source: IEA[a].
[a]*Energy subsidies: tracking the impact of fossil fuel subsidies https://www.iea.org/topics/energy-subsidies.*

devoted to fossil fuel subsidies and Libya, also with 16.7%. Egypt, Algeria, Uzbekistan and Turkmenistan all spend over 5% of GDP on fuel subsidies. The level in China is 0.2% of GDP.

Renewable technologies also receive subsidies, but of a different sort to those most widely used in support of fossil fuel consumption. There has been an effort in many parts of the developed world to encourage the use of renewable electricity generation by the application of incentives of different sorts. For example, the US government has supported wind and solar power through a system of tax credits. Elsewhere, there are feed-in tariffs that allow renewable generators to sell power to the grid at a predetermined price and contracts for difference which make up the payment to a renewable generator so that it achieves a fixed tariff level.

IRENA has recently attempted to estimate in a systematic way the total level of subsidies for renewable generation[1] and how these might evolve over the next 30 years based on a scenario in which the world remains on track to meet the United Nations Framework Convention on Climate Change (UNFCCC) Paris Agreement climate change target of keeping global warming to 2_C or less.

Table 8.3 shows the results of the IRENA analysis, with figures for renewable power generation subsidies for 2017 for several nations and regions as well as an estimate of the global total. The region with the largest level of support for renewable generation is the EU which has set ambitious targets for renewable generation and emissions reduction. IRENA estimates that the EU subsidised renewable generation with around US$78.4bn through feed-in tariffs, green certificates, investment grants and some other tariff support schemes. Within the EU, Germany offered the highest level of support, followed by Italy, the United Kingdom and Spain.

Japan provided US$18.8bn in renewable support in 2017 as the nation seeks to reduce its reliance on imported of fossil fuels; the country has no significant indigenous fossil fuel resources to call upon and imports all its fossil fuels. Japanese support primarily takes the form of feed-in tariffs designed to encourage solar photovoltaic deployment. China also provided around US$15.2bn in 2017 through feed-in tariffs for wind and solar power aimed at accelerating deployment. Meanwhile, the United States subsidised renewable generation with about US$8.9bn through tax credits and investment tax breaks. IRENA found support of around US$2.9bn in India, while for the rest of the world, the cumulative total was US$3.8bn. Based on these figures, the organisation estimated that the global total subsidy for renewable power generation in 2017 was US$128bn.

---

1. Energy Subsidies: Evolution in the Global Energy Transformation to 2050, Michael Taylor, IRENA 2020.

TABLE 8.3 Global subsidies for renewable power generation, 2017.

Country/Region	Subsidy level (US$ billion)
European union	78.4
Japan	18.8
China	15.2
United States	8.9
India	2.9
Rest of the world	3.8
Total	128

Source: IRENA[a].
[a]*Energy Subsidies: Evolution in the Global Energy Transformation to 2050, Michael Taylor, IRENA 2020.*

Broken down by technology, solar photovoltaic received the largest share of subsidies in 2017 around 48% of the total or US$60.8bn. Onshore wind received US$31.6bn or 25% of the total, biomass US$21.9bn (17%) and offshore wind US$6.6bn (5%).

Table 8.4 shows how IRENA has predicted that subsidy regimes will evolve over the coming 30 years, with estimates for subsidy levels in 2030 and 2050 to complement the figures for 2017. The predicted trend is for fossil fuel subsidies to fall sharply between 2017 and 2030 and then continue to tail off towards 2050, while renewable subsidies rise slowly.

Fossil fuel subsidies of US$447bn in 2017 are, on this basis, predicted to fall to US$165bn by 2030, a fall of 63%, and then to US$139bn in 2050 or 69% lower than in 2017. Over the same period, renewable subsidies (unlike

TABLE 8.4 The evolution of energy sector subsidies, 2017–50.

Energy source	2017 (US$ billion)	2030 (US$ billion)	2050 (US$ billion)
Fossil fuel	447	165	139
Nuclear energy	21	27	21
Renewable energy[a]	166	192	209

Source: IRENA[b].
[a]*This figure includes subsidies for transportation fuel as well as power generation. The figure for power generation alone is US$ billion 128.*
[b]*Energy Subsidies: Evolution in the Global Energy Transformation to 2050, Michael Taylor, IRENA 2020.*

Table 8.3, the figures in this table include transportation biofuels as well as subsidies for power generation) rise from US$166bn in 2017 to US$192bn in 2030 and US$209bn in 2050. This last figure is still less than half the level of fossil fuel subsidies in 2017.

Table 8.4 also includes estimates of the level of nuclear power generation subsidies. The organisation suggests that these are much more obscure than the subsidies for fossil fuels and renewables and therefore more difficult to pin down accurately. Subsidies include government support for nuclear waste management and in the case of a new nuclear plant at Hinkley Point in the United Kingdom, significant tariff support of perhaps as much as US$1.4bn/ year.[2] However, the global total for nuclear power is relatively small compared to either fossil fuels or renewables. IRENA estimates that the total was US$21bn in 2017. This is predicted to rise to US$27bn in 2030 but then fall back to US$21bn in 2050.

Externalities

The analyses above indicate the level of direct subsidies for both fossil fuels and renewable generating technologies. However this does not account for all the subsidies because it fails to cost negative externalities, particularly those associated with fossil fuels. These subsidies are unpriced and therefore invisible to consumers except in so far as they have an impact on their lives.

The consumption of fossil fuels leads to an environmental impact that can be both serious and wide ranging. The most significant today is global warming caused by the release of carbon dioxide into the atmosphere when fossil fuels are consumed.[3] There are also much more localised effects on health and hence mortality caused by the emission of harmful pollutants such as particulates from diesel engines, nitrogen oxides from engines and from power plants and sulphur dioxide from coal-fired power stations. These latter can also release heavy metals into the atmosphere.

Estimating the cost of these negative effects on the environment is extremely difficult. The IMF has recently published a study in which is used the concept of an economically efficient fossil fuel price to estimate the level of hidden subsidy associated with fossil fuel consumption.[4] Broadly, an economically efficient price is a price at which the cost of production plus the cost of mitigating any negative effects of the use of the fuel is balanced by the cost the consumer pays for the fuel. The gap between the economically

2. Energy Subsidies: Evolution in the Global Energy Transformation to 2050, Michael Taylor, IRENA 2020.
3. Other greenhouse gases such as methane are released too.
4. IMF Working Paper. Global Fossil Fuel Subsidies Remain Large: An Update Based on Country-Level Estimates, David Coady, Ian Parry, Nghia-Piotr Le and Baoping Shang, IMF, May 2019.

efficient price and the actual price paid by consumers (in this case significantly lower than the economic cost) is the external, unaccounted cost.

Using this approach, the IMF concluded that global energy subsidies were US$4.7 trillion in 2015 and rose to US$5.2 trillion in 2017. Other sources have arrived at different figures. For example, IRENA put the cost of unpriced externalities for fossil fuels at US$3.1 trillion in 2017, lower by US$2.1 trillion but still a staggeringly large figure. The IMF figure indicates that total fossil fuel subsidies are 30 times those received by renewable generating technologies, while the IRENA figure puts the multiple at 19.

This underpricing of the cost of fossil fuel combustion can be broken down into components. According to the IMF, the largest component is underpricing for local air pollution, which accounted for 48% of the estimate in 2015. Global warming accounted for a further 24% and underpricing of the environmental cost of road fuels accounted for a further 15%. This insight suggests that while global warming cannot be controlled by one country alone, local taxes or incentives to reduce pollution levels can have a significant effect locally on air quality and hence human health.

As far as the individual fuels are concerned, coal carries the largest unpriced subsidy, 44% of the total, followed by petroleum with 41% and natural gas with 10% while direct electricity output carries a further 4%. However, given that most of the world's coal is burnt to generate electricity and a large quantity of natural gas is used for power generation too, the subsidy contribution to power generation based on fossil fuels will be more significant than this.

The IMF analysis also breaks down the total subsidy including externalities by country. The top 10 nations by subsidy level are shown in Table 8.5 for 2015. Head of the league, by a large amount, is China. The country, which generates around two-thirds of its electricity from coal, provided an estimated overall subsidy including external costs of US$1432bn. The level of subsidy was more than twice that of the next nation, the United States, with US$649bn. As with China, the United States has relied heavily on coal plants for electricity generation. However, the amount of coal in the US power generation mix has been falling since the beginning of the second decade of the century and this decline continues as natural gas and renewable sources become more important.

Other nations with large total subsidy levels in Table 8.5 include Russia with subsidies of US$551bn, India with US$209bn, Japan with US$177bn and Saudi Arabia with US$117bn. The top 10 nations are rounded off with Iran (US4111bn), Indonesia (US$97bn), Germany (US$72bn) and Turkey (US$64bn).

TABLE 8.5 Ten largest energy subsidies including externalities by country, 2015.

Country	Post tax subsidy (US$ bn)
China	1432
The United States	649
Russia	551
India	209
Japan	177
Saudi Arabia	117
Iran	111
Indonesia	97
Germany	72
Turkey	64

Source: IMF[a].
[a]IMF Working Paper. Global Fossil Fuel Subsidies Remain Large: An Update Based on Country-Level Estimates, David Coady, Ian Parry, Nghia-Piotr Le and Baoping Shang, IMF, May 2019.

Taxes

One way that the imbalance in pricing resulting from externalities can be corrected is by the use of direct taxes or other financial tools. A number of such tools have been developed. These can be applied to correct any of the types of imbalance discussed above but today they are most often used to tackle the issue of global warming and the release of carbon dioxide and other greenhouse gases into the atmosphere as a result of fossil fuel combustion. These measures are variously known as carbon taxes or carbon pricing mechanisms.

The WB has identified five types of initiatives that attempt to put a price on greenhouse gas emissions.[5] The first and simplest is a carbon tax. This sets a fixed price that must be paid for the release of 1 tonne of carbon dioxide equivalent (tCO_2equivalent) into the atmosphere. The tax might be framed as an excise duty or a levy but it is essentially a carbon tax.

The second financial instrument is an emissions trading system (ETS). These come in two forms. A cap-and-trade system sets an annual cap on the total quantity of greenhouse gas that can be released into the atmosphere within a particular political region and then issues a set number of emissions

---

5. State and Trends of Carbon Pricing 2020, World Bank Group, May 2020.

certificates equivalent to this amount each year, each one permitting to the release of 1 tCO_2equivalent of these gases. These certificates may be allotted to particular emitters or they may be auctioned. Any facility emitting greenhouse gases must then submit a certificate for each tCO_2equivalent that it releases. However organisations with certificates can also sell them on the ETS market where other organisations might choose to buy them in order to increase the quantity they can emit. An alternative system, called baseline-and-credit, sets a baseline level of emissions that each regulated emitter can release. If the emitter does not reach this baseline, it can be issued with certificates for the difference which it can sell on the ETS market. Emitters which seek to exceed their baseline must buy and surrender certificates for all their excess emissions.

A third type of financial instrument is called a carbon crediting mechanism. This allows a jurisdiction to issue certificates for projects that actively reduce emissions beyond any regulated level. This might involve financing the planting of trees or supporting renewable development in another country. Any certificates issued in this way can then be traded for financial gain. Finally, result-based climate finance is a system whereby targets are set, and upon reaching the target, a recipient will receive funds from the finance provider.

The most important global carbon emission schemes in operation are based either on carbon taxes or on ETS systems. The WB analysis revealed that there were 61 carbon pricing initiatives in place or scheduled for implementation at the end of 2019. These included 31 ETS systems and 30 carbon taxes. It estimated that these covered around 22% of global greenhouse gas emissions. Together they allowed governments to raise over US$45bn in 2019.

One of the largest schemes is the European Union's ETS scheme, which was launched in 2005. The scheme is based on the European Union Allowance or EUA. One EUA allows the holder the right to emit 1 tonne of carbon dioxide or the equivalent for N_2O and perfluorocarbons. The market price of a trading certificate has varied markedly since the scheme was launched. The prices of units traded in the early years, when units were allocated, were relatively high but they stabilised towards the end of the first decade of the century. Table 8.6 shows the price over the 10 years between January 2010 and January 2020. In 2010, the cost was €12.79, and in 2011, it had risen to €14.28. The cost fell back after that, to €6.35 in 2012 and as low as €4.59 in 2014. The cost remained relatively low until 2018 when prices started to rise sharply so that at the beginning of 2019 the market price was €22.24 and in 2020 it reached €24.26.

In this, and other similar schemes, the unit price is intended to act as a market signal that will influence emitters. If the cost to emit greenhouse gases is too high, then consumers will switch to alternative energy sources, but if it is

TABLE 8.6 Cost of EU ETS carbon unit 2010–20.

Year	EUA (€)[a]
2010	12.79
2011	14.28
2012	6.35
2013	6.10
2014	4.59
2015	6.82
2016	7.77
2017	5.52
2018	7.78
2019	22.24
2020	24.26

Source: Ember[b].
[a]*The cost is for the first Monday in January of each year.*
[b]*Ember Carbon Price Viewer https://ember-climate.org/data/.*

too low, then it may be cheaper to pay the price and continue to emit. The IEA believes that a carbon price of US$75–100/tCO$_2$equivalent is needed to maintain a trajectory that would keep the world in line with the commitments in the Paris Agreement on climate change. On the other hand, the IMF has suggested that some countries can meet their Paris Agreement targets with a price in 2030 of US$35.[6] Others would need at least double of that to achieve the same end.

In addition to this and other ETS schemes, there are a number of countries that have introduced carbon taxes. The highest carbon tax is found in Sweden where the unit cost is US$119/tCO$_2$equivalent. Switzerland and Liechtenstein have taxes of US$99/tCO$_2$equivalent. However almost half of the emissions that are subject to pricing have a cost of less than US$10/tCO$_2$equivalent. This will not provide a strong enough signal to encourage the change of behaviour needed to combat global warming.

---

6. Putting a Price on Pollution, Ian Parry, Finance and Development, December 2019.

Similar incentives and systems have been used to control other harmful emissions from power plants. For example, the United States introduced a cap-and-trade system in the 1990s to control the emission of sulphur dioxide from coal-fired power plants. A scheme to limit nitrogen oxide emissions was introduced in 2003.

As already noted, rather than distorting the cost of electricity, all these schemes attempt to price in the external cost of the targetted emissions. However, their imposition is generally a political decision.

Chapter 9

The cost of electricity

The cost of electricity is important at every level of global electricity systems. Consumers will normally seek the lowest cost supplier, commensurate with their needs. In a deregulated electricity, market retail and wholesale suppliers will seek to buy their electricity from the lowest cost generator. Meanwhile, generators will seek the lowest cost source for the power they intend to generate. And while a range of factors will come to bear of the final cost of a unit of electricity, the overriding factor will generally be the cost of electricity from a power station.

The electricity industry is a conservative industry based on production units — power plants — that are expected to last for years if not decades. Even so, there is constant change. Demand rises as societies advance. Those societies begin to demand cleaner power. And power plants get old and must eventually be retired. Each of these factors can lead to the need for new generating capacity to be built.

The typical lifetime of many types of generating plant is around 30 years, so even without considering other factors, capacity must be replaced or old plants rehabilitated over this timescale. And each time a new power plant is required, a decision must be taken about the type of power plant to be built. Sometimes this decision will be based on local factors such as a particular resource that can usefully be exploited, but in virtually all cases, one of the key considerations will be the type of power plant that can provide the cheapest electricity.

The cost of electricity from a power plant of any type depends on a range of factors. First, there is the cost of building the power station and buying all the components needed for its construction. In addition, most large power projects today are financed using loans, so there will be a cost associated with paying back the loan, with interest. Then there is the cost of operating and maintaining the plant over its lifetime, including fuel costs. Finally, the overall equation should include the cost of decommissioning the power station once it is removed from service.

It would be possible to add up all these cost elements to provide a total cost of building and running the power station over its lifetime, including the cost of decommissioning, and then divide this total by the total number of units of electricity that the power station produced over its lifetime. The result would be the real lifetime cost of electricity from the plant. Unfortunately such

The Cost of Electricity. https://doi.org/10.1016/B978-0-12-823855-4.00009-7

calculation could only be completed once the power station was no longer in service. From a practical point of view, this would not be of much use. The point in time at which the cost-of-electricity calculation of this type is most needed is before the power station is built. This is when a decision is made to build a particular type of power plant.

Levelized Cost of energy model

In order to get around this problem, economists have devised a model that provides an estimate of the lifetime cost of electricity before the station is built. Of course, because the plant does not yet exist, the model requires a large number of assumptions to be made. In order to make this model as useful as possible, all future costs are converted to the equivalent cost today by using a parameter known as the discount rate. The discount rate is almost the same as the interest rate and relates to the way in which the value of one unit of currency falls (most usually, but it could rise) in the future. This allows, for example, the maintenance cost of a steam turbine 20 years into the future to be converted into an equivalent cost today. The discount rate can be applied to the cost of electricity from each type of plant in 20 years' time too.

The economic model is called the levelized cost of electricity (LCOE) model, a type of lifecycle analysis that was discussed briefly in Chapter 6. The model contains a lot of assumptions and flaws, but it is the most commonly used method available for estimating the cost of electricity from a new power plant.

The LCOE model treats all types of power station equally. However, there are significant differences between technologies that must also be taken into account. A major division is between the main combustion technologies and nuclear power, on the one hand — technologies where a fuel is required to maintain output — and, on the other hand, the primary renewable technologies, wind, solar and hydropower, which exploit a free renewable resource.

One difference between the groups relates specifically to the energy source. Power stations based on combustion technology such as coal-fired and gas-fired power plants, as well as nuclear plants, all require a fuel to be supplied continuously in order to operate and this fuel comes at a regular cost. This type of plant can often be relatively cheap to build, so the cost of the fuel will play a large role in determining the cost of electricity from each station. Plants based on the main renewable technologies may be more expensive to build, but there are no fuel charges and so the cost of electricity from these plants is very closely related to the relative cost of building each type of plant.

Another important difference between the two types of technology is what is known as dispatchability. Power plants based on combustion and nuclear technologies can be controlled to deliver power as needed; they are considered dispatchable. The renewable technologies are intermittent and unpredictable and they cannot be relied upon to provide power when required; these are usually considered non-dispatchable and this affects the market value of the electricity they produce.

In order to take account of this difference, the US Energy Information Administration (US EIA), which assembles LCOE tables for each type of technology for its *Annual Energy Outlook* each year, has in recent years begun to add a new type of economic modelling called the levelized avoided cost of electricity (LACE). This tries to take account of the different levels of grid service each type of technology provides. Combining the LCOE and the LACE can provide a fuller picture of the benefits of each technology.

The LCOE estimates the future cost of electricity from a particular technology. Looked at another way, this figure may also be considered to be the cost that must be charged for electricity from the plant if it is to cover the cost of its construction and operation. The LCAE, in contrast, is an estimate of the revenue that the new plant would be able to expect from its electricity in the prevailing market if it were constructed. This will depend on the competing plants that are available to supply power at the same time. In principle, therefore, if the LCOE is higher than the LACE, then the plant will be operating at a loss while if the LACE is higher than the LCOE, the plant would be economically attractive to build.

The LCOE remains the simplest and most used metric for comparing the generating costs of different technologies and that is the model that is used for the most part in what follows. When it is used, its limitations should be kept in mind.

Electricity generating costs

The LCOE from different types of generating plant will vary from place to place. Many of the figures quoted below are based on the US market but even here there can be wide variations depending upon location. The main factors that will lead to differences are the variable cost of fuels in different places and to a smaller extent the variation in the cost of labour required to build and operate power plants. The figures quoted below are mostly what are known as overnight costs which do not include any financing costs. The latter must be added to provide the real cost in any specific situation but for the purposes of comparison the overnight cost is a more valuable metric.

The financial advisory and asset management company Lazard has been publishing an annual levelized cost analysis of the power sector in the United States for over a decade. Figures from the 2019 report are shown in Table 9.1. The analysis includes calculation of the LCOE for new power stations from a range of conventional and renewable technologies, in most cases providing a range of final costs, the variation reflecting the difference found in the cost of electricity from the same power source in different parts of the United States. This may be considered a proxy for the variability that is likely to be found in other parts of the world — although this assumption should not be applied too loosely. The United States is, after all, still the richest nation on earth.

TABLE 9.1 Levelized cost of electricity for US generating technologies (2019).

Technology	Cost ($/MWh)
Coal with carbon capture and storage	66–152
Natural gas combined cycle	44–68
Open cycle gas turbine	150–199
Nuclear	118–192
Utility-scale solar photovoltaic (PV)	36–44
Commercial and industrial rooftop solar PV	75–154
Domestic rooftop solar PV	151–252
Solar thermal with energy storage	126–156
Onshore wind	28–54
Offshore wind	89
Geothermal	69–112

Source: Lazard.[a]
[a]*Lazard's Levelized Cost of Energy Analysis - Version 13.0, Lazard.*

Based on the Lazard analysis, the new fossil fuel fire plant with the lowest cost electricity in 2019 was a natural gas-fired combined cycle plant with an LCOE of US$44−68/MWh. The natural gas plant is not fitted with carbon capture and storage (CCS). For a coal-fired power station *with* CCS, the estimated LCOE is US$66−152/MWh, while the LCOE for a natural gas-fired open cycle gas turbine was US$150−199/MWh. Power from the latter units is expensive, but they are generally only used to provide power to the grid during periods of peak demand. Meanwhile, the cost of power from a new nuclear power plant was estimated to be US$118−192/MWh. On this basis, nuclear power looks like an uneconomical choice for new technology compared to either coal or natural gas.

It is notable, however, based on the figures in Table 9.1, that the cheapest new source of electricity in the United States is from none of these plants because they are all undercut by the best renewable technologies. The most competitive is onshore wind power which has an LCOE of US$28−54/MWh, followed by utility-scale solar photovoltaic (solar PV) with a levelized cost of US$36−44/MWh. This calculation does not take account of the cost of grid support needed for renewable generation, but it does provide a useful guide to the effectiveness of these technologies.

Offshore wind is more expensive than onshore wind, with an estimated LCOE of US$89/MWh. Two other solar PV costs are also included in the table. The calculated LCOE for commercial and industrial rooftop solar PV was US$75−154/MWh, significantly higher than for utility-scale solar PV. And domestic rooftop solar PV with an LCOE of US$151−252/MWh was more expensive still. However it is important to remember that rooftop solar PV is a distributed technology that feeds power either directly to the consumer or into a distribution grid. At this point in the electricity system, power is much more expensive than at the transmission system level. So, even power as expensive as this can still be competitive.

A figure for solar thermal generation, US$125−156/MWh, is also included in Table 9.1. This refers to a plant with energy storage which is considerably more dispatchable than solar PV. The cost of new geothermal power is also included, with an estimated LCOE of US$69−112/MWh. There is limited geothermal capacity available, anywhere in the world.

The US EIA produces an annual estimate of the LCOE from a range of technologies as part of its *Annual Energy Outlook*. Figures from the most recent report are shown in Table 9.2. The EIA takes a slightly different approach to LCOE analysis by providing an estimate for the cost of electricity from different technologies all entering service at the same future date. As some plants, a nuclear power plant for example, may take 5 years from initial order to entering service, the estimates in the first column of Table 9.2 are for plants entering service in 2025.

As with the data in Table 9.1, the fossil fuel station that will provide the lowest cost power in 2025 is a natural gas-fired combined cycle plant with an LCOE of US$38/MWh. For an ultra-supercritical coal-fired power plant, this time without CCS, the estimated cost of power is US$76/MWh. Both are broadly in line with the earlier table. However, the cost of power from an open cycle gas turbine, US$67/MWh, is notably lower than in the previous table as is the cost of electricity from an advanced nuclear power plant at US$82/MWh. These differences reflect differences in the assumptions made about these two technologies in the two studies.

Again the US EIA analysis reveals that the main renewable technologies are extremely competitive based on the LCOE analysis. Solar PV is the cheapest source in Table 9.2 with an estimated LCOE of US$36/MWh while onshore wind is close behind at US$40/MWh. The electricity from a geothermal power plant was estimated by the US EIA to be much cheaper than in the previous table, at US$37/MWh. Offshore wind remains expensive with an LCOE for plants entering service in 2025 of US$122/MWh. Table 9.2 also contains an estimated LCOE for a hydropower plant entering service in 2025. At US$53/MWh, this is relatively competitive. However, a new biomass power plant, with an estimated LCOE of US$95/MWh in 2025, looks relatively expensive.

TABLE 9.2 Levelized cost of electricity for US generating technologies, 2025 and 2040.

Technology	Entering service in 2025 (US$/MWh)	Entering service in 2040 (US$/MWh)
Ultra-supercritical coal	76	72
Natural gas combined cycle	38	43
Open cycle gas turbine	67	72
Advanced nuclear	82	74
Geothermal	37	37
Biomass	95	87
Onshore wind	40	36
Offshore wind	122	86
Solar photovoltaic	36	30
Hydropower	53	54

Source: US EIA.[a]
[a]*Levelized Cost and Levelized Avoided Cost of New Generation Resources in the Annual Energy Outlook 2020, US Energy Information Administration, 2020.*

In addition to the LCOE for power plants entering service in 2025, the US EIA has also calculated the LCOE for the same type of plant entering service in 2040. These figures are shown in the second column of Table 9.2. Most of the changes from one column to the next are small. For a coal-fired power plant, the LCOE drops from US$76/MWh in 2025 to US$72/MWh in 2040, while for a natural gas-fired combined cycle plant, it rises from US$38/MWh to US$43/MWh. The cost of power from an advanced nuclear plant falls from US$82/MWh to US$74/MWh in 2040. The cost of biomass power stays the same while that for hydropower increases very slightly.

The cost of electricity from the main renewable sources apart from hydropower falls between 2025 and 2040. For solar PV, the LCOE in 2025 of US$36/MWh drops to US$30/MWh in 2040. The LCOE for onshore wind falls from US$40/MWh to US$36/MWh and for offshore wind the cost falls from US$112/MWh to US$86/MWh. These figures imply a further improvement in the competitiveness of renewable power compared to fossil fuel and nuclear as the century progresses.

Table 9.3 compares the LCOE for the main generating technologies with the LACE for plants entering service in 2025. Again these figures are from the

TABLE 9.3 Comparison of LCOE and LACE for US generating technologies entering service in 2025.

Technology	LCOE (US$/MWh)	LACE (US$/MWh)
Ultra-supercritical coal	76	36
Natural gas combined cycle	38	37
Advanced nuclear	82	36
Biomass	95	42
Onshore wind	40	32
Offshore wind	122	34
Solar photovoltaic	36	34
Hydropower	53	35

US EIA. As noted earlier, a higher LACE should imply that a plant based on the technology will be economically viable while a higher LCOE suggests that the full costs of operating the plant will not be met. However, none of the estimates for the LACE in Table 9.3 are higher than the corresponding LCOE. It is also notable that the LACE figures in the table occupy a narrow band of costs, from US$32/MWh to US$42/MWh.

There are, nevertheless some significant differences between the sets of figures in the two columns of Table 9.3. For a natural gas-fired combined cycle plant, the LCOE and LACE are US$38/MWh and US$37, respectively, while for solar PV, the two figures are US$36/MWh and US$34/MWh; in both cases, the difference is probably too small to be significant. For onshore wind, the LCOE is US$40/MWh and the LACE US$32/MWh, putting new wind power at a slightly greater disadvantage than either solar PV or a combined cycle plant. However for coal-fired technology the LACE is US$36/MWh while that of LCOE is US$76/MWh and for an advanced nuclear plant the LACE is US$36/MWh while the LCOE is US$82/MWh. These figures suggest both would operate at a significant loss based on this simple comparison. A similar conclusion applies to biomass and offshore wind while hydropower with an LOCE of US$53/MWh in 2025 and an LACE of US$35/MWh falls in the middle ground between best and worst.

There are no comprehensive sets of global estimates of the cost of electricity from the complete range of different generating technologies to set against these US figures, but the International Renewable Energy Agency (IRENA) has published costs for the main renewable generating technologies for the decade from 2010 to 2019. These are shown in Table 9.4.

TABLE 9.4 Global average annual LCOE for renewable technologies ($/MWh).

Year	Solar photovoltaic	Solar thermal	Onshore wind	Offshore wind	Hydropower	Bioenergy
2010	378	346	86	161	37	76
2011	286	348	83	175	36	55
2012	223	353	83	154	38	61
2013	175	268	82	177	43	81
2014	164	243	76	183	44	82
2015	126	251	69	169	39	73
2016	114	290	66	146	52	72
2017	92	253	64	131	55	72
2018	79	184	58	127	45	57
2019	68	182	53	115	47	66

Source: IRENA.[a]
[a]*Renewable power Generation Costs in 2019, IRENA.*

The figures in the last row of Table 9.4 can be compared with those in Table 9.1 in order to gauge in some measure where the US market fits into the global market. The global average LCOE for solar PV in 2019 from Table 9.4 is US$68/MWh, well above the US$36—44/MWh in the United States. The average global cost of a new solar thermal power plant in 2019, US$182/MWh, is also higher than the US estimate of US$126—156/MWh. On the other hand, onshore wind with a global average LCOE of US$53/MWh does fall within the range found in the United States of US$28—54/MWh. However, the LCOE for offshore wind power, US$115/MWh, is higher than the estimate for the United States of US$89/MWh.

Hydropower does not appear in the Lazard analysis and in the figures in Table 9.1. Table 9.4 has a global average LCOE for this technology, US$47/MWh. This is the cheapest source of power in the table, outperforming both onshore wind and solar PV. The United States and the developed nations of Europe have exploited their best hydropower sites so there is little scope for large expansion here but there remain good resources to exploit in other parts of the world.

The figures in Table 9.4 also show the cost trends for these technologies. The trends for the individual technologies will be examined in more detail below, but the salient feature of this table is the fall in the cost of electricity from solar PV plants over the decade noted in the table. The LCOE in 2010 was over five times higher than in 2019. There is a fall in the LCOE for onshore wind too, but it is much smaller. The following sections will look in more detail at trends in the LCOE for the technologies in Table 9.1.

Coal-fired power plants

Table 9.5 shows figures for the LCOE of electricity from a new coal-fired power plant with CCS based on analysis from Lazard between 2009 and 2019. In this case the table contains a single figure representing the average price across the United States each year rather than a range as in Table 9.1 for a single year, 2019.

The LCOE figures in the table, with two exceptions, show little variation. In 2009, Lazard estimated an LCOE of US$123/MWh, and in 2011, it estimated a cost of US$95/MWh. Otherwise the figures all fall between US$102/MWh and US$109/MWh. Costs rose slightly during the middle of the decade shown and then fell back before rising again at the end of the decade.

Coal-fired power generation is becoming increasingly unpopular, particularly across the developed world, as a result of the high greenhouse gas emissions from coal combustion. Its use in the United States has been declining since the beginning of the second decade. However coal continues to be popular in some developing nations, particularly in India and China. The figures in Table 9.5 are for a plant with CCS and this puts the technology at a disadvantage compared to a combined cycle plant without CCS or compared to

TABLE 9.5 Levelized cost of electricity (LCOE) for US coal-fired plants.

Year	LCOE ($/MWh)
2009	123
2010	107
2011	95
2012	102
2013	105
2014	109
2015	108
2016	102
2017	102
2018	102
2019	109

Source: Lazard.[a]
[a]*Lazard's Levelized Cost of Energy Analysis - Version 13.0, Lazard.*

onshore wind and solar PV. It should be noted, however, that new coal-fired power plants are not being built with CCS and so their costs are likely to be lower.

The International Energy Agency (IEA), Nuclear Energy Agency (NEA) and Organisation for Economic Co-operation and Development (OECD) have produced a series of five yearly reports called *Projected Cost of Generating Electricity* (IEA report). The eighth of these reports was published in 2015. The report uses a different methodology to that for the reports so far cited so the results are not directly comparable. In particular, the LCOE figures include financing costs at a range of discount rates. The LCOE estimate also includes a carbon cost of US$30/t carbon dioxide.

The IEA report estimated the LCOE for an advanced coal-fired plant in the United States without CCS operating at a capacity factor of 85% as between US$83/MWh (discount rate 3%) and US$104/MWh (discount rate 10%). In China, the range was US$74/MWh−US$82/MWh while in Japan the LCOE range was US$95/MWh−US$119/MWh.

Coal-fired power plants are generally designed for base load operation at their maximum capacity factor. If these plants are operated at a lower capacity factor efficiency is likely to fall and emissions may rise. In consequence the LCOE rises for a capacity factor of 50% instead of 85%. In the United States, for example, with a 50% capacity factor, the range of LCOEs was US$101/MWh−US$137/MWh. The increase in the use of renewable generation means

TABLE 9.6 Levelized cost of electricity (LCOE) for US natural gas—fired combined cycle plants.

Year	LCOE ($/MWh)
2009	83
2010	96
2011	95
2012	75
2013	74
2014	74
2015	64
2016	63
2017	60
2018	58
2019	56

Source: Lazard.[a]
[a]*Lazard's Levelized Cost of Energy Analysis - Version 13.0, Lazard.*

that many former base load plants such as coal-fired power plants are being required to operate at a lower capacity factor than in the past so the sensitivity of the LCOE to capacity factor should be taken into account today.

Natural gas—fired combined cycle plants

The LCOE for a US natural gas-fired combined cycle power plant between 2009 and 2019 is shown in Table 9.6. Unlike the figures for coal-fired plants, above, there is (excepting the first figure[1]) a monotonic decrease in the cost over the period. In 2010, the estimated LCOE was US$96/MWh but by 2015 this had fallen to US$64/MWh and in 2019 the LCOE was US$56/MWh, a fall of 42% over 9 years. The cost of electricity from a natural gas combined cycle power plant is very sensitive to the cost of natural gas and in the United States, during the decade from 2010, the cost of natural gas has been falling as a consequence of the development of shale gas deposits in the country. A similar LCOE price trend may not, therefore, be found in other parts of the world.

---

1. Several of the LCOE figures for 2009 from the Lazard report appear anomalous and may suggest a chance in assumptions between 2009 and 2010. Alternatively this may be a result of a change in conditions resulting from the global financial crisis of 2007—08.

Figures from the IEA for the cost of electricity[2] in the United States in its 2015 report provide an LCOE of US$61/MWh at a discount rate of 3% rising to US$71/MWh at a discount rate of 10%. This is broadly in line with Table 9.6 when the additional carbon cost used by the IEA is taken into account. Elsewhere, the LCOE varies markedly. In the United Kingdom, for example, the LCOE range was US$213/MWh (3%) to US$263/MWh (10%) for the most expensive plant cited in the report. In China, the cost varied between US$90/MWh and US$96/MWh.

Capacity factor also has an effect on the cost of power from a combined cycle power plant. The US cost from a plant at a 3% discount rate and 85% capacity factor of US$61/MWh rose to US$68/MWh at 50% capacity factor. Modern combined cycle plants are relatively capable when it comes to modulating their output but efficiency falls and emissions can rise as the capacity factor falls.

Open cycle gas turbine plants

The open cycle gas turbine is an agile, fast-acting power unit that can be brought into service rapidly and removed again swiftly. This has made it one of the main sources of peak power on grids across the world. The units tend to be relatively expensive and less efficient than the base load plants such as the combined cycle plant, and this is reflected in the LCOE of power from these power units.

Table 9.7 shows the LCOE trend for open cycle gas turbines according to the annual Lazard analysis. From an LCOE of US$275/MWh in 2009 in the United States the cost fell to US$192/MWh in 2015 and US$175/MWh in 2019. As with the natural gas-fired combined cycle plant discussed above, the fall in the cost of electricity from these plants is mostly attributable to the fall in the cost of natural gas in the United States. The IEA does not consider open cycle gas turbines in its five yearly reports, but the LCOE from these units in other parts of the world will likely be higher than in the United States.

Nuclear power plants

Nuclear power is a controversial technology that is being promoted as a carbon free source of power in some constituencies but at the same time is being phased out elsewhere. While old nuclear power plants that have had their construction costs paid down can be a cheap source of power, the economic argument revolves around the cost of new nuclear power. Table 9.8 presents Lazard figures for the levelized cost of new nuclear power in the United States between 2009 and 2019. In 2009, the estimated LCOE was US$123/MWh. This fell sharply in 2010 to US$96/MWh but by 2015, it had risen to US$117/

---

2. Projected Costs of Generating Electricity: 2015 edition, IEA, 2015.

TABLE 9.7 Levelized cost of electricity (LCOE) for US open cycle gas turbine.

Year	LCOE ($/MWh)
2009	275
2010	243
2011	227
2012	216
2013	205
2014	205
2015	192
2016	191
2017	183
2018	179
2019	175

Source: Lazard.[a]
[a]Lazard's Levelized Cost of Energy Analysis - Version 13.0, Lazard.

TABLE 9.8 Levelized cost of electricity (LCOE) for US nuclear power plants.

Year	LCOE ($/MWh)
2009	123
2010	96
2011	95
2012	96
2013	105
2014	112
2015	117
2016	117
2017	148
2018	151
2019	155

Source: Lazard.[a]
[a]Lazard's Levelized Cost of Energy Analysis - Version 13.0, Lazard.

MWh and in 2019 it was US$155/MWh. This puts nuclear power at a significant disadvantage compared to both coal-fired technology and natural gas–fired combined cycle technology. The rise in LCOE reflects a steep rise in the capital cost of nuclear power in the United States.

The IEA in its report of 2015 estimated the LCOE of nuclear power in the United States to be US$54/MWh at a 3% discount rate and US$102/MWh at a 10% discount rate. Costs for new plants in Europe were comparable. However for two Asian nations the costs were estimated to be much lower. In South Korea, the estimated LCOE in 2015 was US$27/MWh (3%) to US$51/MWh (10%), while in China the estimated cost was US$26/MWh to US$49/MWh for the cheapest plant cited. The stark variation in cost between the United States and Europe on the one hand and South Korea and China on the other is the result of the difference in capital costs. These are substantially lower in both China and South Korea.

Nuclear power plant costs are also sensitive to capacity factor. Nuclear plants have traditionally been considered as base load generators and they do not usually operate comfortably at low capacity factors. According to IEA estimates, the LCOE for power from a US nuclear power plant operating in 2015 rose from US$54/MWh at an 85% capacity factor to US$77/MWh at a capacity factor of 50%. This increase is typical of European nuclear plants too, but the Asian plants (China and South Korea) were estimated to be less sensitive.

Onshore wind power plants

The LCOE for onshore wind power plants in the United States between 2009 and 2019 is shown in Table 9.9. The figures in this table show a sharp discontinuity between 2010 and 2011. In 2009, the estimated LCOE was US$135/MWh, and in 2010, it was US$124/MWh but in 2011 it had fallen to US$71/MWh. After that the LCOE falls, a trend consistent with the gradual improvement in wind power capital cost and performance over the period. In 2015 the estimated LCOE was US$55/MWh and by 2019 it had fallen to US$41/MWh, a fall of 42% in 8 years.

The IEA 5-year report shows a wide variability in the cost of wind power in the United States in 2015. The LCOE from the lowest cost plant cited was estimated to be US$33/MWh (3% discount rate), while the most expensive was US$116/MWh. (As a comparison, the Lazard analysis from 2015 showed an overnight LCOE range of US$32–77/MWh.[3]) At a discount rate of 10%, the IEA range was US$52–188/MWh. Similar variability was reported by the IEA elsewhere. In South Korea, for example, the LCOE for wind power was as high as US$214/MWh (3%). Meanwhile, in China, the lowest LCOE was US$46/MWh (3%), the next lowest cost after the USA figure.

---

3. Lazard's Levelized Cost of Energy Analysis - Version 9.0, Lazard, November 2015.

TABLE 9.9 Levelized cost of electricity (LCOE) for US onshore wind power plants.

Year	LCOE ($/MWh)
2009	135
2010	124
2011	71
2012	72
2013	70
2014	59
2015	55
2016	47
2017	45
2018	42
2019	41

Source: Lazard.[a]
[a]*Lazard's Levelized Cost of Energy Analysis - Version 13.0, Lazard.*

As the cost of wind power scales closely to the capital cost of building wind power plants, this variation between and within nations reflects a similar variation in capital costs.

The cost of electricity from wind turbines does not vary with their capacity factor. However wind power is both intermittent and unpredictable and this means that while it will usually be dispatched when available, there must always be a source to replace it when the wind fails. In consequence, while the cost of electricity from wind power plants can be extremely competitive, it is considered less valuable to the grid.

Offshore wind farms are more expensive to develop than similar facilities onshore. Against this, they usually offer a better wind regime and plants of large aggregate capacity can be developed. As with onshore wind, the LCOE of electricity from these plants will scale with the capital cost of their development. For example, Bloomberg New Energy Finance estimated the average global LCOE for offshore wind in 2019 to be US$78/MWh.[4] Most offshore wind capacity is located in European waters.

---

4. New Energy Outlook 2019, Bloomberg New Energy Finance, 2019. Figures are taken from Latest BNEF Report Finds Levelized Cost Of Renewables Continues To Fall, Steve Hanley, CleanTechnica, 30 October 2019.

TABLE 9.10 Levelized cost of electricity (LCOE) for US utility-scale crystalline solar photovoltaic power plants.

Year	LCOE ($/MWh)
2009	359
2010	248
2011	157
2012	125
2013	104
2014	79
2015	65
2016	55
2017	50
2018	43
2019	40

Source: Lazard.[a]
[a]Lazard's Levelized Cost of Energy Analysis - Version 13.0, Lazard.

Solar photovoltaic power plants

The cost of electricity from solar PV power plants has shown the most dramatic change over the 10 years from 2009 to 2019 of any generating technology. This was highlighted in Table 9.4 and is emphasised again with the figures from Lazard in Table 9.10. In 2009, the estimated average LCOE from US utility solar PV plants was US$359/MWh, the highest of all the main power generating technologies. This fell sharply in the succeeding years, to US$248/MWh in 2010 and US$157/MWh in 2011. By 2015, the estimated average LCOE for US solar PV was US$65/MWh, and in 2019, it had fallen to US$40/MWh, making it the least cost source of all the primary technologies.

This dramatic fall in the cost of solar electricity is a result of a massive fall in the cost of solar cells. During the decade from 2009, these devices became global commodities with cells manufactured in China particularly competitive. While the rate at which costs are falling has slowed, there may still be room for a further decrease.

The LCOE figures in Table 9.10 are for utility-scale solar PV plants. A large proportion of solar PV installations are on rooftops. These tend to be of smaller capacity than the utility plants and the costs are significantly higher. For example, the LCOE estimates from the 2019 Lazard analysis puts the cost of a commercial and industrial rooftop facility at around twice the cost of a utility-scale installation. A domestic rooftop PV installation is around twice as much again. At the same time, the cost of electricity to these consumers is much higher than the wholesale price of power. For example, in the United States, in June 2019, the average retail cost of electricity to a US consumer was US$134/MW while for a commercial customer it was US$109/MWh.[5] With consumer prices at these levels, rooftop solar PV installations can provide competitively prices electricity.

The IEA 5-year report provides LCOE estimates from 2015 for a range of solar PV installations across the globe.[6] For a large US utility-scale plant, the LCOE was US$54/MWh at 3% discount rate and US$103/MWh at a 10% discount rate. The Chinese LCOE was similar at US$55/MWh (3%) and US$87/MWh (10%). In comparison, the LCOE for a French plant of US$104/MWh was high, as it was in South Korea at US$102/MWh, both at 3% discount rate.

For domestic rooftop installations, the LCOE in the United States was US$106/MWh (3%) while in France it was US$214/MWh and in South Korea it was US$156/MWh. The wide variation in prices reflects the differing local markets for solar cells. It should also be remembered that the steep fall in the cost of solar cells means that prices for all these categories at the beginning of the third decade of the 21st century are likely to be much lower.

Solar thermal power plants

Solar thermal power plants are hybrid power generators that utilise the heat from the sun to drive a thermodynamic engine, usually a steam turbine, to provide electricity. As such they are much more complex than solar PV plants and capital costs are significantly higher.

Table 9.11 presents figures from Lazard for the LCOE for a US solar thermal power plant between 2009 and 2019. The figures in the table show no consistent trend. The estimated cost of electricity in 2009 was US$168/MWh. This had fallen to US$124/MWh by 2014, but the cost rose in succeeding years and then fell again, so that in 2019 the LCOE was US$141/MWh.

These figures suggest that solar thermal electricity is relatively expensive. However, a solar thermal plant can include energy storage, allowing the plant to supply electricity at night as well as during the day. This makes the power from these plants more readily dispatchable and therefore more valuable.

The IEA 5-year report from 2015 contains some LCOE estimates for solar thermal power plants. In the United States, the LCOE for a solar thermal plant with 6h of energy storage was US$79/MWh (3% discount rate) while for a plant with 12h or energy storage it was US$66/MWh. For a plant with storage in South Africa (storage capacity not quoted), the LCOE was US$139/MWh,

5. Electric Power Monthly, US Energy Information Administration, June 2020.
6. Projected Costs of Generating Electricity: 2015 edition, IEA, 2015.

TABLE 9.11 Levelized cost of electricity (LCOE) for US solar thermal power plants (solar tower).

Year	LCOE ($/MWh)
2009	168
2010	157
2011	159
2012	174
2013	145
2014	124
2015	150
2016	151
2017	140
2018	140
2019	141

Source: Lazard.[a]
[a]*Lazard's Levelized Cost of Energy Analysis - Version 13.0, Lazard.*

while in Spain for a plant without storage capacity it was US$263/MWh. This is a developing technology and prices may fall much lower if the global installation volume was to rise. However, these plants are currently only chosen in exceptional circumstances.

Geothermal power plants

Geothermal power plants are relatively cheap to build, but the cost of prospecting for underground geothermal reservoirs suitable to power a generating plant usually increases the capital cost of development and this affects the LCOE from such plants. On the other hand if a resource, once found, is managed well then once capital costs are paid down these plants become a cheap source of electricity.

Table 9.12 contains LCOE estimates for new geothermal power plants in the United States between 2009 and 2019. The first figure in the table appears anomalous. The LCOE for a US geothermal plant in 2010 was estimated to be US$107/MWh. This rose to US$116/MWh in 2012 but started to fall in 2015 and by 2019 the LCOE was estimated to be US$91/MWh. These estimates make geothermal power relatively expensive compared with other renewable technologies. Nevertheless, where geothermal reserves suitable for power generation exist, it has often proved economical to exploit.

TABLE 9.12 Levelized cost of electricity (LCOE) for US geothermal power plants.

Year	LCOE ($/MWh)
2009	76
2010	107
2011	104
2012	116
2013	116
2014	116
2015	100
2016	98
2017	97
2018	91
2019	91

Source: Lazard.[a]
[a]Lazard's Levelized Cost of Energy Analysis — Version 13.0, Lazard.

The 2015 IEA report on the *Projected Costs of Generating Electricity* contains a small number of LCOE estimates for geothermal electricity in countries that have useful geothermal resources. In the United States, the LCOE was US$55/MWh at 3% discount rate and US$99/MWh at a discount rate of 10%. Turkey is another country that has exploited geothermal energy. The LCOE range there from the 2015 report was US$109/MWh (3%) to US$123/MWh (10%). In Italy, the range was US$60/MWh—US$100/MWh.

Other generating technologies

The cost of electricity from hydropower plants depends on the capital cost, which is usually relatively high. Against that, many hydropower plants have extremely long lifetimes and once their capital outlay is paid off the power they generate is at a low cost. The 2015 IEA report contains some LCOE estimates for hydropower. These show a wide variation. In Portugal, for example, the LCOE for power from a 144 MW dam and reservoir plant was estimated to be US$90/MWh at a 3% discount rate, rising to US$284/MWh at a discount rate of 10%. In Turkey, the LCOE for power from a 20 MW hydropower plant was US$30/MWh at 3% discount rate and US$54/MWh at 10%. For a 1000 MW pumped storage project in Switzerland, the equivalent costs were US$36/MWh (3%) and US$107/MWh (10%). These figures can be compared with the figures in Table 9.4, which show the global average LCOE for hydropower in 2015 to be US$39/MWh (but in 2016, it was US$55/MWh).

The LCOE for a 100 MW biomass plant in the United States, from the IEA report of 2015,[7] was US$99/MWh at 3% discount rate and US$138/MWh at 10%. For a 10 MW plant in Spain, the LCOE was US$152/MWh (3%) and US$190/MWh (10%). Biomass is a combustion technology but the plants are small compared with typical fossil fuel plants and efficiencies are much lower. This leads to higher capital costs, reflected in these figures.

The wide variation in the cost of electricity from some of the technologies discussed above makes it important to take into account important regional factors such as variations in the cost of loans, different local legislation and the different cost of commodities and labour in different countries and regions. The LCOE provides a convenient metric for comparing the future cost of electricity but it must be used wisely.

---

7. Projected Costs of Generating Electricity: 2015 edition, IEA, 2015.

Index

Printed in the United States
By Bookmasters